Nanostructure Science and Technology

Series Editor

David J. Lockwood, FRSC
National Research Council of Canada
Ottawa, ON, Canada

Nanostructure science and technology now forms a common thread that runs through all physical and materials sciences and is emerging in industrial applications as nanotechnology. The breadth of the subject material is demonstrated by the fact that it covers and intertwines many of the traditional areas of physics, chemistry, biology, and medicine. Within each main topic in this field there can be many subfields. For example, the electrical properties of nanostructured materials is a topic that can cover electron transport in semiconductor quantum dots, self-assembled molecular nanostructures, carbon nanotubes, chemically tailored hybrid magnetic-semiconductor nanostructures, colloidal quantum dots, nanostructured superconductors, nanocrystalline electronic junctions, etc. Obviously, no one book can cope with such a diversity of subject matter. The nanostructured material system is, however, of increasing significance in our technology-dominated economy and this suggests the need for a series of books to cover recent developments.

The scope of the series is designed to cover as much of the subject matter as possible – from physics and chemistry to biology and medicine, and from basic science to applications. At present, the most significant subject areas are concentrated in basic science and mainly within physics and chemistry, but as time goes by more importance will inevitably be given to subjects in applied science and will also include biology and medicine. The series will naturally accommodate this flow of developments in the sciences and technology of nanostructures and maintain its topicality by virtue of its broad emphasis. It is important that emerging areas in the biological and medical sciences, for example, not be ignored as, despite their diversity, developments in this field are often interlinked. The series will maintain the required cohesiveness from a judicious mix of edited volumes and monographs that while covering subfields in depth will also contain more general and interdisciplinary texts.

Thus the series is planned to cover in a coherent fashion the developments in basic research from the distinct viewpoints of physics, chemistry, biology, and materials science and also the engineering technologies emerging from this research. Each volume will also reflect this flow from science to technology. As time goes by, the earlier series volumes will then serve as reference texts to subsequent volumes.

Ramachandra Naik · H. P. Nagaswarupa ·
H. C. Ananda Murthy · Mika E. T. Sillanpää
Editors

Advanced Nanomaterials for Energy Storage Devices

Springer

Editors
Ramachandra Naik
Department of Applied Sciences
New Horizon College of Engineering
Bengaluru, Karnataka, India

H. C. Ananda Murthy
Department of Applied Sciences
Papua New Guinea University
of Technology
Lae, Papua New Guinea

H. P. Nagaswarupa
Department of Studies in Chemistry
Davangere University
Davangere, Karnataka, India

Mika E. T. Sillanpää
Department of Biological and Chemical
Engineering
Aarhus University
Aarhus C, Denmark

ISSN 1571-5744 ISSN 2197-7976 (electronic)
Nanostructure Science and Technology
ISBN 978-3-031-74729-8 ISBN 978-3-031-74730-4 (eBook)
https://doi.org/10.1007/978-3-031-74730-4

This Springer imprint is published by the registered company Springer Nature Switzerland AG
The registered company address is: Gewerbestrasse 11, 6330 Cham, Switzerland

Preface

The field of energy storage is undergoing a profound transformation, driven largely by the advent of nanomaterials. These materials, with their unique properties and potential to revolutionize energy storage systems, are at the forefront of current research and technological innovation. As global energy demands rise and the transition to sustainable energy sources becomes increasingly urgent, the need for advanced materials that can enhance the efficiency, capacity, and longevity of energy storage devices has never been more critical.

This book, *Advanced Nanomaterials for Energy Storage Devices*, is a comprehensive exploration of the latest advancements in the design, synthesis, characterization, and application of nanomaterials in energy storage. It is intended to serve as both a reference and a guide for researchers, engineers, and students who are engaged in this dynamic and rapidly evolving field.

The book begins with Chap. 1, "Introduction to Nanomaterials for Energy Storage", setting the stage by discussing the fundamental concepts and the transformative potential of nanomaterials in improving energy storage technologies. Following this, Chap. 2 delves into the "Fundamentals of Computational Design in Nanomaterials", providing readers with the theoretical foundation needed to understand and design nanomaterials at the atomic level.

Building on this foundation, Chap. 3, Computational Modeling of Energy Storage Mechanisms, explores the role of computational techniques in understanding and optimizing the behavior of nanomaterials in energy storage applications. This chapter, revised under the title "Computational Modeling of Radiative Magneto Maxwell Nanofluid: An Application to Improving the Efficiency of Energy Conversion and Storage Systems", emphasizes the intersection of computational science and material engineering in the quest for more efficient energy solutions.

The subsequent chapters focus on the practical aspects of nanomaterial research. Chapter 4, "Characterization Methods for Nanomaterials in Energy Storage" provides a detailed overview of the tools and techniques used to analyze nanomaterials, while Chap. 5, "Synthesis Techniques for Novel Nanomaterials" and Chap. 6, "Advanced Synthesis Strategies for Enhanced Energy Storage Performance" explore the cutting-edge methods used to create materials with specific properties tailored to energy storage applications.

In situ characterization is an emerging area of interest, and this is thoroughly covered in Chap. 7, titled "In-Situ Characterization Techniques for Energy Storage Applications", where the dynamic behavior of nanomaterials within operational energy storage devices is examined. To bridge theory with practice, Chap. 8 titled "Case Studies: Nanomaterials in Specific Energy Storage Devices" presents real-world applications, showcasing the impact of these advanced materials in various energy storage systems.

The book also addresses the critical aspect of electrode development in Chap. 9, "Advanced Electrode for Energy Storage: Types and Fabrication Techniques", which highlights the latest innovations in electrode materials and fabrication processes.

The final chapters of the book, Chap. 10, "Challenges and Opportunities in Nanomaterials for Energy Storage" and Chap. 11, "Future Perspectives and Emerging Trends in Nanomaterials for Energy Storage", provide a thoughtful reflection on the current state of the field, discussing the hurdles that remain, as well as the promising new directions that research is likely to take in the coming years.

Writing this book has been a deeply enriching experience. The intersection of nanotechnology and energy storage is a fertile ground for discovery, and it is my hope that this work will inspire further innovation and exploration. The challenges we face in energy storage are complex, but with the continued development of advanced nanomaterials, the solutions are within reach.

I would like to express my gratitude to the many colleagues, researchers, and students whose insights and contributions have enriched this work. Their dedication to advancing the field has been a constant source of inspiration.

Finally, I hope that this book serves as a valuable resource for those engaged in the exciting and critical work of developing the next generation of energy storage devices. The future of energy is being shaped by the discoveries of today, and it is my sincere belief that the advancements discussed in this book will play a pivotal role in that future.

Bengaluru, India — Ramachandra Naik
Davangere, India — H. P. Nagaswarupa
Lae, Papua New Guinea — H. C. Ananda Murthy
Aarhus C, Denmark — Mika E. T. Sillanpää

Contents

1 **Introduction to Nanomaterials for Energy Storage** 1
D. M. Tejashwini, H. P. Nagaswarupa, Ramachandra Naik, and A. S. Manjunatha

2 **Fundamentals of Computational Design in Nanomaterials** 25
S. Sreejith, J. Ajayan, N. V. Uma Reddy, J. M. Radhika, Jimsha K. Mathew, B. Sivasankari, and N. V. Raghavendra Reddy

3 **Computational Modeling of Radiative Magneto Maxwell Nanofluid: An Application to Improving the Efficiency of Energy Conversion and Storage Systems** 41
Pudhari Srilatha, M. C. Jayaprakash, Konduru Sarada, R. J. Punith Gowda, and R. Naveen Kumar

4 **Characterization Methods for Nanomaterials in Energy Storage** 57
B. C. Vasantha Kumar, R. Madhu Kumar, H. P. Shivarudrappa, B. M. Santhosh, and P. N. Prashantha Kumar

5 **Synthesis Techniques for Novel Nanomaterials** 77
Abbay Gebretsadik Debalke, Buzuayehu Abebe, and H. C. Ananda Murthy

6 **Advanced Synthesis Strategies for Enhanced Energy Storage Performance** 99
Jayanti Mishra, Arabinda Baruah, and Jyotirekha Dutta

7 **In-Situ Characterization Techniques for Energy Storage Applications** 111
Yashwanth Venkatraman Naik and D. M. Tejashwini

8 **Case Studies: Nanomaterials in Specific Energy Storage Devices** 123
B. M. Santhosh, H. P. Shivarudrappa, B. C. Vasantha Kumar, P. N. Prashantha Kumar, Soniya D. Naik, Prasannakumar Jammapra Kallappa, and T. Divya

9 **Advanced Electrode for Energy Storage: Types and Fabrication Techniques** 139
Solomon Jenoris Muthiya, M. Rudresh, G. S. Divya, B. E. Naveena, and Mohankumar Subramanian

10 **Challenges and Opportunities in Nanomaterials for Energy Storage** 165
D. M. Tejashwini, H. M. Deepa, V. V. Deshmukh, and Yashwanth Venkatraman Naik

11 **Future Perspectives and Emerging Trends in Nanomaterials for Energy Storage** 183
Yashwanth Venkatraman Naik

Index 197

Chapter 1
Introduction to Nanomaterials for Energy Storage

D. M. Tejashwini, H. P. Nagaswarupa, Ramachandra Naik, and A. S. Manjunatha

Abstract The chapter titled "Introduction to Nanomaterials for Energy Storage" delves into the transformative role of nanomaterials in enhancing the performance of energy storage systems, such as batteries, supercapacitors, and fuel cells. By leveraging nanoscale properties—such as increased surface area and enhanced catalytic activity—nanomaterials have significantly improved energy density, charge/discharge rates, and cycle life. This chapter explores various types of nanomaterials, including metal oxides, sulphides, carbon-based materials, and nanocomposites, along with their synthesis methods like hydrothermal synthesis and sol–gel techniques. Advanced characterization tools, such as Raman spectroscopy, XRD, TEM, and SEM, are discussed in the context of their roles in elucidating nanostructures. While the chapter acknowledges challenges related to environmental impact, scalability, and integration, it also highlights the potential of emerging materials, such as hybrid nanomaterials and 2D structures. These innovations offer promising avenues for sustainable energy solutions, paving the way for a new era of energy storage technologies that are more efficient, sustainable, and capable of meeting global energy demands.

Keywords Nanomaterials · Energy storage

D. M. Tejashwini · H. P. Nagaswarupa (✉)
Department of Studies in Chemistry, Shivagangothri Davangere University, Davangere 577007, India
e-mail: nswarupa@davangereuniversity.ac.in

R. Naik
Department of Physics, New Horizon College of Engineering, Bangalore 560103, India

A. S. Manjunatha
Department of Chemistry, Don Bosco Institute of Technology, Bengaluru, Karnataka 560074, India

R. Naik et al. (eds.), *Advanced Nanomaterials for Energy Storage Devices*, Nanostructure Science and Technology, https://doi.org/10.1007/978-3-031-74730-4_1

1.1 Introduction

1.1.1 Definition of Nanomaterials

Materials with structural elements that are less than 100 nanometres (nm) in at least one dimension are referred to be nanomaterials. Their size gives rise to their special characteristics, enabling them to show higher surface area-to-volume ratios and quantum effects when compared to bulk materials. There are several different kinds of nanomaterials, such as nanoparticles, nanowires, nanotubes, and nanosheets [1]. Nanoparticles can have dramatically different qualities, such as increased mechanical strength, electrical conductivity, or chemical reactivity. Their large surface area, which promotes interactions with the surroundings, is the cause of this. Many disciplines, including medical, electronics, energy, and environmental research, have found uses for nanomaterials [2, 3]. Their capacity to interact at the cellular level with biological systems makes their development and use subject to rigorous examination of safety and environmental consequences [4].

Quantum effects and higher surface area-to-volume ratios give them special qualities, which radically change their chemical, biological, and physical behaviour. Materials frequently display quantum confinement at the nanoscale, which modifies their optical and electrical characteristics. As an illustration, semiconductor nanoparticles exhibit size-dependent band gaps that are tunable for certain optical uses [2]. Quantum dots are used in cutting-edge imaging and display technologies. Compared to bulk materials, nanomaterials have a substantially higher surface area per unit volume. As a result of their increased surface area, they are more reactive and interact with other chemicals, which makes them excellent catalysts [1]. Nanocatalysts use this feature to facilitate industrial processes and environmental applications.

Owing to their smaller size, nanomaterials frequently show higher mechanical strength. Carbon nanotubes have remarkable elasticity and tensile strength, which is advantageous for creating high-performance structural components and reinforcing composite materials [4]. Nanomaterials electrical and thermal conductivities can vary significantly from those of their bulk counterparts. The high conductivity of nanowires and nanotubes can be advantageous in heat management and nanoelectronics applications [5]. When nanoparticles interact with light, they can produce special optical phenomena like surface plasmon resonance. Utilizing this characteristic improves sensitivity and resolution in biosensing and imaging applications [6]. The magnetic properties of nanomaterials might vary depending on their size. Superparamagnetism may be seen in magnetic nanoparticles, which is advantageous for targeted drug delivery systems and biomedical imaging [7]. Nanomaterials are extremely important and adaptable in many sectors, such as electronics, health, energy, and environmental research, because of their special characteristics at the nanoscale.

1.1.2 Importance of Energy Storage

Energy storage plays a vital role in contemporary energy systems, facilitating effective energy management, distribution, and utilization. It is essential for tackling the problems of incorporating renewable energy sources, enhancing grid stability, and satisfying the world's expanding energy needs. This chapter explores the importance of energy storage, the range of technologies it uses, and how it affects many industries. Solar and wind energy are instances of inconsistent renewable energy sources that can provide electricity. Systems for storing surplus energy produced during periods of peak production and discharging it during periods of low output or high demand are known as energy storage systems (ESS) [8]. This feature facilitates the integration of renewable energy sources into the grid by increasing their stability and dependability.

Energy storage systems provide functions including voltage support, spinning reserve, and frequency management, all of which have a major positive impact on grid stability. They can react swiftly to variations in the supply and demand of energy, preserving the electrical grid's dependability. To provide a steady supply of electricity and prevent blackouts, this skill is crucial [9, 10]. Communities and countries may improve their energy security and lessen their reliance on foreign fuels by storing energy locally. In times of emergency or natural catastrophe, energy storage devices may boost resilience and self-sufficiency by supplying backup power [11]. Lithium-ion batteries, flow batteries, and solid-state batteries are just a few examples of the energy storage technologies that have recently advanced and seen considerable cost and performance improvements. These technologies are well suited for a variety of applications, ranging from large-scale grid storage to consumer devices, thanks to their high energy density, extended cycle life, and enhanced safety [12].

A growing number of current technologies and environmental initiatives depend on nanomaterials because of their special qualities at the nanoscale. Their uses are numerous and range from improving environmental preservation and energy storage to transforming the fields of electronics and medical technology. The efficiency of energy storage devices like batteries and supercapacitors is being greatly enhanced using nanomaterials. They are perfect for increasing energy density, charging rates, and overall efficiency because of their high surface area-to-volume ratio, electrical conductivity, and mechanical strength [13]. Compared to standard materials, nanostructured electrodes in lithium-ion batteries can enhance capacity and longevity. Similar to this, supercapacitors employ nanomaterials like carbon nanotubes and graphene to increase their energy and power densities [14].

Nanomaterials play an important part in environmental sustainability by assisting with pollution management and reconstruction. Nanoparticles like titanium dioxide (TiO_2) are used in photocatalysis to remove heavy metals from polluted water and to break down organic contaminants. Water and air pollutants may be effectively treated by nanomaterials due to their great selectivity and reactivity [15]. Electronics is being revolutionized by nanomaterials, which make it possible to create quicker, smaller, and more efficient devices. Graphene, nanowires, and quantum dots are essential materials for high-performance transistors, memory chips, and sensors. These materials boost computing power, enable miniaturization, and enhance electronic device utility [16].

Nanomaterials are revolutionizing medication delivery, diagnosis, and therapy in the medical field. By modifying them to target certain cells or tissues, nanoparticles enhance the accuracy of treatments and diagnostic imaging. Liposomal nanoparticles improve medication delivery efficiency, whereas gold nanoparticles are employed in targeted cancer therapy and imaging. Additionally, nanomaterials are improving renewable energy technologies like hydrogen generation and solar cells. Photovoltaic cells are made more efficient by quantum dots and nanostructured materials, which increase charge carrier mobility and light absorption. Hydrogen synthesis from water is a crucial step in creating sustainable energy and uses nanocatalysts [17, 18].

Energy-efficient coatings and thermal insulation are two of the many areas where nanomaterials can save energy. The use of nanostructured materials in coatings and insulation improves building energy conservation and lowers total energy usage. Innovations in energy storage, electronics, health care, environmental clean-up, and renewable energy technologies are propelled by their special qualities. The incorporation of nanoparticles holds the potential for further enhancing technical capabilities and supporting sustainable development goals as research continues to progress [19].

1.2 Overview of Energy Storage Technologies

1.2.1 Lithium-Ion Batteries

Lithium-ion (Li-ion) batteries are a cornerstone of modern energy storage technology as shown in Fig. 1.1, with high energy density, extended cycle life, and a low self-discharge rate making them popular in consumer devices, electric vehicles (EVs), and grid energy storage. These batteries are renowned for their dependability and efficiency and use lithium compounds as their main source of charge carriers. Three basic parts make up a standard Li-ion battery: an electrolyte, a negative electrode (anode), and a positive electrode (cathode). Graphite or lithium titanate are frequently used for the anode, whereas lithium metal oxides (such as $LiCoO_2$ and $LiFePO_4$) are typically used for the cathode. During charge and discharge cycles, the electrolyte typically a lithium salt mixed in an organic solvent helps transfer lithium ions between the anode and cathode [20, 21].

Li-ion batteries have several benefits, one of which is their high energy density, which permits more energy storage in a less amount of space. In addition, they have a high life cycle ranging hundreds to thousands of charge–discharge cycles and a low self-discharge rate, which keeps the battery charged while not in use. These characteristics make them perfect for applications including renewable energy, EVs, and portable electronics. Li-ion technology has advanced recently with an emphasis

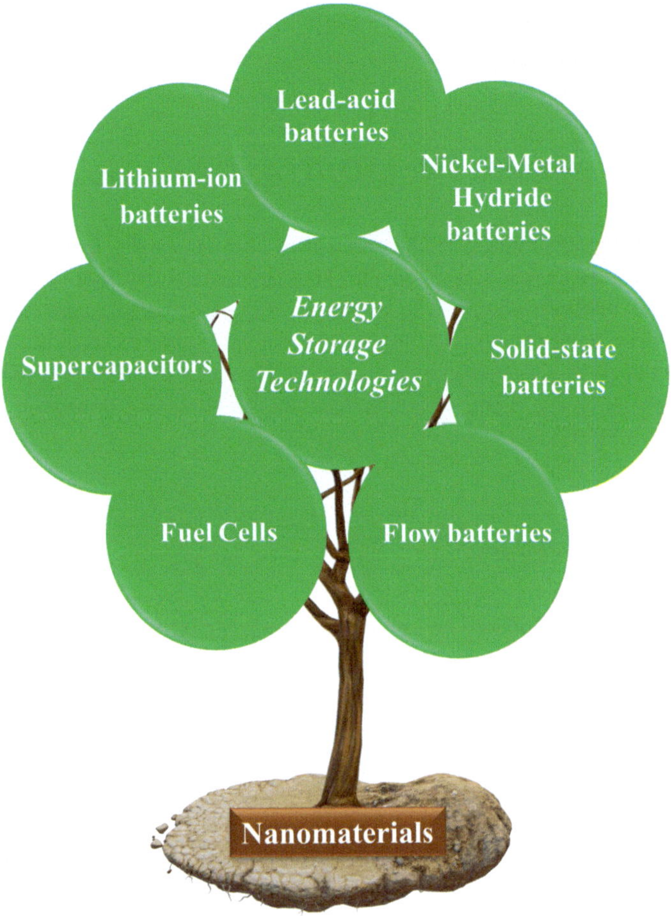

Fig. 1.1 Different energy storage technologies

on cost reduction, energy density increase, and safety enhancement. One example of such innovation is the creation of solid-state batteries, which increase energy density and safety by using a solid electrolyte rather than a liquid. To further improve battery performance, researchers are also looking at novel cathode and anode materials, such as silicon-based anodes [22].

Apart from lithium-ion batteries, other noteworthy battery varieties are lead-acid and nickel-metal hydride (NiMH) batteries, each with unique attributes and uses. Different energy storage technologies are illustrated in Fig. 1.1.

Lead-acid batteries are one of the oldest forms of rechargeable batteries. They are made up of sulphuric acid (H_2SO_4) as the electrolyte, sponge lead (Pb) for the negative plate, and lead dioxide (PbO_2) for the positive plate. These batteries are preferred due to their affordability and dependability, which makes them perfect for backup power and automotive applications. However, they have important disadvantages like as poor energy density and short cycle life. If recycled improperly, the lead concentration also raises environmental problems [23].

1.2.2 Nickel-Metal Hydride (NiMH) Batteries

NiMH batteries have a hydrogen-absorbing alloy as the negative electrode and nickel oxide hydroxide as the positive electrode. They are extensively utilized in consumer gadgets and hybrid cars and provide a better energy density than lead-acid batteries. Because NiMH batteries do not contain harmful heavy metals, they offer a better environmental profile than lead-acid batteries. They can experience self-discharge problems, which cause them to rapidly lose charge while not in use, and they are more expensive [24].

1.2.3 Supercapacitors

Supercapacitors are a kind of energy storage technology that fills the void left by batteries and regular capacitors. They are sometimes referred to as ultracapacitors or electrochemical capacitors. In contrast to batteries, supercapacitors may charge and discharge far more quickly because they store energy via electrostatic fields rather than chemical processes. They are made up of a separator and an electrolyte between two electrodes. Activated carbon and other materials with a high surface area are frequently used to make electrodes, which increases capacitance [25]. Compared to conventional batteries, supercapacitors provide a number of benefits, such as a high-power density, quick cycles of charging and discharging, and extended operating lifetimes [26]. Supercapacitors, on the other hand, store less energy per unit volume or mass than batteries because they have a lower energy density. This means that rather than being useful for long-term energy storage, they can only be used in situations where high power and fast charge/discharge cycles are essential.

1.2.4 Fuel Cells

Fuel cells are electrochemical devices that have a minimal environmental effect and a high efficiency of converting chemical energy into electrical energy. Their working principle is the combination of an oxidant (generally airborne oxygen) with a fuel (usually hydrogen) to generate heat, water, and power. An electrolyte separates the fuel and oxidant in an electrochemical cell, where this process takes place. The most popular fuel cell types are solid oxide fuel cells (SOFCs), which are renowned for their high efficiency and versatility in fuel types, alkaline fuel cells (AFCs), which are highly efficient and found in space missions, and proton exchange membrane (PEM) fuel cells, which are widely used in automotive applications due to their high-power density and quick start-up times [27]. Fuel cells have various benefits, including high efficiency, lower greenhouse gas emissions, and the possibility for renewable energy integration. Nevertheless, they encounter difficulties, such as exorbitant expenses, problems with storing hydrogen, and inadequate infrastructure for fuel delivery. The main goals of ongoing research include lowering prices, increasing fuel cell durability, and creating effective technologies for producing and storing hydrogen [28].

1.2.5 Emerging Technologies

Solid-state batteries are a significant improvement in energy storage technology, providing greater safety, energy density, and longevity than typical lithium-ion batteries. Solid-state batteries employ a solid electrolyte as opposed to conventional batteries, which use liquid or gel electrolytes. By removing the flammability hazards connected to liquid electrolytes and lowering the possibility of leaks, this modification improves safety. The increased energy density of solid-state batteries is one of their main advantages. Batteries with solid electrolytes can store more energy per unit weight or volume because they can support greater energy densities. This is especially helpful for applications where weight and space are crucial, such as electric cars and portable gadgets. Furthermore, solid-state batteries often have longer cycle lives and superior thermal stability, making them more durable over extended use. In order to make solid-state batteries economically feasible, current research is concentrated on increasing the ionic conductivity of solid electrolytes, creating novel materials, and streamlining production procedures [29, 30].

The capacity of flow batteries to store and release electrical energy through electrochemical processes in liquid electrolytes makes them a potential energy storage technology. The electrolytes in these batteries are kept in two different tanks and are pumped through a cell stack where the electrochemical processes take place. Flow batteries are especially well suited for grid storage and large-scale applications because of their architecture, which permits adjustable system scaling and scalable energy storage. The extended cycle life and excellent efficiency of flow batteries are two of their main benefits [31, 32]. They provide a dependable answer for energy storage requirements, withstanding several charge–discharge cycles without experiencing appreciable deterioration. Moreover, the flexibility of flow batteries makes them useful for large-scale applications as it makes it simple to modify the storage capacity simply adjusting the size of the electrolyte tanks, concentrating on creating novel electrolyte materials and streamlining system design in order to increase the flow batteries' efficacy and affordability [33].

1.3 Role of Nanomaterials in Energy Storage

1.3.1 Enhancing Energy Density

Nanomaterials considerably increase energy density in energy storage systems through the employment of their unique nanoscale characteristics. Nanomaterials like carbon nanotubes and silicon nanoparticles increase the electrode efficiency and storage capacity in batteries by providing greater surface areas and quicker electron/ion transport channels. Nanomaterials, such as graphene and metal oxides, enhance the surface area that may be utilized for storing charge, thereby elevating the energy density of supercapacitors. Nanomaterials also help to accelerate charge/discharge rates and improve cycle stability. Studies have indicated that the integration of nanoparticles into electrode materials can result in significant enhancements to specific energy and power density [34]. When compared to traditional graphite anodes, silicon anodes with nanostructures, for example, show much better energy densities.

1.3.2 Improving Charge/Discharge Rates

Nanomaterials accelerates ion and electron mobility, increasing charge and discharge rates in energy storage devices. They may diffuse ions more quickly and have better electrode conductivity due to their large surface area-to-volume ratio. Shorter paths for charge carriers in batteries are made possible by nanostructured materials like nanowires and nanosheets, which accelerate charge and discharge cycles. Higher power densities and quicker reaction times are made possible in supercapacitors by the increased electrode surface area caused by nanomaterials, such as graphene and carbon nanotubes. Furthermore, internal resistance is decreased with the aid of nanomaterials, speeding up the charge and discharge operations [35].

1.3.3 Increasing Cycle Life

The life cycle of energy storage devices can be greatly extended by nanomaterials because they improve structural stability and lessen deterioration during charge and discharge cycles. Because they can accommodate expansion and contraction more easily than other materials, nanostructured materials, such as nanoparticles and nanocomposites, may aid in preventing the mechanical breakdown of electrode materials [36]. Nanoengineered silicon-based anodes exhibit better capacity retention and less volume growth than their bulk counterparts. Furthermore, by enhancing the electrode-electrolyte contact, nanomaterials might lessen side reactions which frequently result in capacity loss over time. Nanomaterials, such as graphene and carbon nanotubes, are used in supercapacitors to prevent considerable material loss and to retain excellent performance over a long period of cycles, hence improving the electrodes' endurance and lifetime [37].

1.3.4 Reducing Costs

With ability to enhance the functionality and efficiency of numerous components of energy storage systems, nanomaterials are essential in lowering their cost. The energy and power density of batteries and supercapacitors are frequently increased by nanostructured electrodes, enabling the creation of smaller, more effective devices with reduced material requirements. Production expenses are decreased by this decrease in material utilization. Additionally, over time, cheaper replacement and maintenance costs may result from the improved performance and extended lifespan of energy storage systems utilizing nanomaterials. Developments in nanomaterials, such as scalable manufacturing methods and economical synthesis strategies, also help to bring down total prices [38]. For instance, advances in manufacturing techniques are making it possible to make nanocarbon materials like carbon nanotubes and graphene at a decreasing cost [39].

1.4 Types of Nanomaterials Used in Energy Storage

1.4.1 Carbon-Based Nanomaterials

Different kinds of nanomaterials are used in energy storage as shown in Fig. 1.2. Graphene, a single sheet of carbon atoms organized in a hexagonal lattice, is a key material in energy storage technologies due to its extraordinary capabilities. Due to its high electrical conductivity, batteries and supercapacitors can charge and discharge at faster rates, which improves efficiency and performance. Graphene is perfect for high-energy density applications owing to its enormous surface area, which enables a larger charge storage capacity. Moreover, the mechanical strength of graphene prolongs the lifespan of energy storage devices by enhancing their stability and endurance. Graphene's large surface area in supercapacitors enables quick charge build-up and discharge, producing high-power density devices. Graphene may be utilized to enhance electrode materials in batteries, increasing their power and energy densities [40].

Carbon atoms organized in a hexagonal lattice form the cylindrical nanostructures known as carbon nanotubes (CNTs). They have a great deal of surface area, strong mechanical strength, and exceptional electrical conductivity, all of which contribute to their effectiveness in energy storage applications. Due to their large surface area and superior electrical conductivity, CNTs increase the charge storage

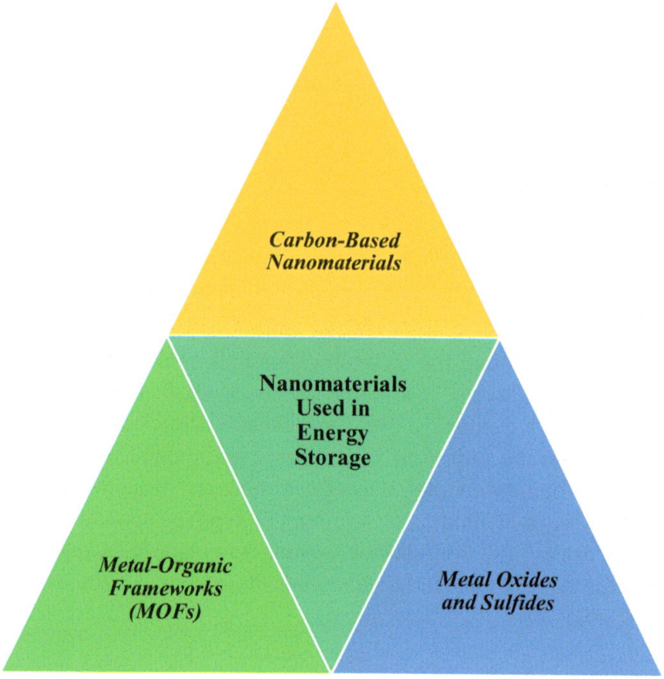

Fig. 1.2 Types of nanomaterials used in energy storage

capacity and charge/discharge rates in supercapacitors. The energy storage devices' endurance and structural integrity are enhanced by their mechanical characteristics. CNTs are utilized to strengthen electrode materials in lithium-ion batteries, hence boosting energy and power densities. Better electron and ion movement is made possible by CNT's capacity to form networks, which is essential for high-performance batteries [41].

Fullerenes are spherical carbon molecules with special electrical characteristics that make them useful for energy storage applications. Examples of these molecules are carbon nanotubes and buckyballs (C_{60}). Their ability to build solid structures and large surface area make them beneficial for improving the performance of batteries and supercapacitors. Fullerenes may be utilized to make high-surface-area electrodes for supercapacitors, which greatly increases charge storage and discharge rates. Their strong electrical conductivity speeds up charge transfer, which improves performance. To further maximize the performance of supercapacitors, fullerenes can also be chemically or doped to improve their interaction with electrolytes [42]. Fullerenes can be included in the anode or cathode materials of lithium-ion batteries to enhance electrical conductivity and raise the battery's energy density. Due to their special molecular arrangement, stable and reversible electrochemical interfaces may form, which is essential for preserving cycle stability and high capacity [43].

1.4.2 Metal Oxides and Sulphides

Transition metal oxides, including titanium dioxide (TiO_2) and manganese dioxide (MnO_2), are essential to energy storage systems owing to their remarkable electrochemical characteristics. Due to its excellent electrochemical stability and high specific capacitance, MnO_2 is commonly used in supercapacitors and batteries. Its quick redox kinetics improve charge/discharge rates and the overall efficiency of the energy storage system. TiO_2 is a typical anode material in lithium-ion batteries because of its stability and safety. The batteries cycle life and safety are enhanced by its sturdy construction and high-rate capabilities. These metal oxides' special qualities, such as their large surface area and strong electrical conductivity, have a major impact on the development of energy storage technologies [44].

1.4.3 Transition Metal Sulphides

Excellent conductivity and electrochemical performance of transition metal sulphides, including molybdenum disulphide (MoS_2), are making them important materials for energy storage applications. In particular, MoS_2 has attracted interest due to its layered structure, which enables exceptional cycle stability and a large charge storage capacity. Due to its special qualities, this material may be used in lithium-ion batteries and supercapacitors. MoS_2 is an appealing option for improving energy storage devices because of its high specific capacitance and advantageous electrochemical profile. Through its efficient ion intercalation, energy storage devices operate more efficiently overall and at higher charge/discharge rates [45].

1.4.4 Metal–Organic Frameworks (MOFs)

Metal-organic frameworks (MOFs) are a type of material that has shown great promise in energy storage due to its unique features. These frameworks have a high surface area and are porous due to the coordination of metal ions or clusters with organic ligands. Their capacity to store and release energy is improved by their large surface area. MOFs are used in energy storage in a variety of ways. Increased ion and electron mobility are made possible by their large surface area, which is essential for high-capacity energy storage. Furthermore, as MOFs are adjustable, their chemical composition and pore diameters may be changed to maximize performance for particular uses [46].

1.4.5 Nanocomposites

Nanocomposites are developed by combining materials at the nanoscale with other substances to improve their characteristics for particular uses, including energy storage systems. Nanocomposites are nanostructured, and they often have a greater surface area, which means that more active sites for electrochemical reactions. Composites with improved performance properties are produced by combining nanomaterials, like metal oxides or sulphides, with carbon-based materials or conductive polymers. The electrical conductivity of composites made of metal oxides (MnO_2, TiO_2) or sulphides (MoS_2) is greatly enhanced by the inclusion of conductive nanomaterials, such as carbon nanotubes or graphene. Faster rates of charge and discharge for batteries and supercapacitors are supported by this improvement. Electrode deterioration during charge–discharge cycles is less common in nanocomposites, which frequently exhibit improved mechanical stability and durability. Investigations have indicated that notable enhancements in energy density and cycle life may be attained by combining metal oxide nanocomposites with graphene. Similarly, because of their large surface area and excellent conductivity, metal sulphide-based nanocomposites provide improved performance in supercapacitors [47].

1.5 Nanomaterial Synthesis and Characterization

1.5.1 Synthesis Methods

1.5.1.1 Chemical Vapour Deposition (CVD)

A flexible and popular method for creating thin films and nanostructures with exact control over composition, structure, and morphology is chemical vapour deposition (CVD). In CVD, a thin film or nanostructure is formed when precursor gases react with a heated substrate surface. In order to affect the development and characteristics of the deposited material, the process takes place in a tightly regulated environment with carefully regulated temperature, pressure, and gas flow rates. Precursor molecules are added to a reactor chamber during CVD, where they undergo chemical reactions that can be aided by plasma or heat. A solid substance is deposited on the substrate surface as a result of these processes. The temperature of the substrate is a major factor in affecting the crystallinity, growth rate, and orientation of material. CVD is critical to many different sectors and scientific domains. It is extensively employed in the production of semiconductors to deposit thin layers of materials like silicon nitride and silicon dioxide. CVD is used in nanotechnology to create carbon-based compounds, such as carbon nanotubes and graphene. Additionally, it is employed in the deposition of thin films of nitrides, metal oxides, and other useful materials utilized in energy storage devices, optics, and electronics. The method has several benefits, including the capacity to cover intricate and irregularly shaped surfaces, precise control over film composition and thickness, and scalability for large-scale manufacturing [48].

1.5.1.2 Sol–Gel Process

The sol-gel technique is a flexible approach for creating metal oxide materials and composites from liquid precursors. Using chemical processes like hydrolysis and condensation, a sol, or colloidal suspension of nanoparticles, is transformed into a gel, or three-dimensional network, in this procedure. At relatively low temperatures, the method makes it possible to produce thin films, coatings, and powders with regulated microstructures. Metal chlorides or alkoxides are commonly employed as precursors in the sol-gel process. In the presence of water, they hydrolyse to produce metal hydroxides, which condense to form a gel network. The final oxide substance is obtained by annealing and drying the gel. The capacity of this process to yield homogenous materials with excellent purity and homogeneity makes it valuable. It is extensively utilized in the production of nanomaterials for implemented for the applications like energy storage, optics, and catalysis [49].

1.5.1.3 Hydrothermal Synthesis

The process of hydrothermal synthesis involves creating a variety of crystalline minerals and nanomaterials in an aqueous solution at regulated pressure and temperature levels. By heating precursor materials to high temperatures, they are placed in a sealed reactor vessel with solvent as reducer and heated. The reactor's high temperature and pressure encourage the development, nucleation, and dissolution of crystals and nanoparticles. The capacity of hydrothermal synthesis to provide nanomaterials with precisely specified sizes, forms, and structures is one of its main benefits. The qualities of the final product are influenced by reaction factors including pH, pressure, and temperature, which may be precisely controlled using this approach. The synthesis of metal oxides, sulphides, hydroxides, and other complicated materials is frequently done using hydrothermal synthesis [50].

1.5.1.4 Electrospinning

A flexible method for creating nanofibers from a polymer melt or solution is electrospinning. A charged jet of polymer is released from a needle or spinneret and moves towards a grounded collection when a strong electric field is applied to a polymer solution or melt. The ultrafine fibres are gathered as a nonwoven mat or film as the solvent evaporates during the process. This approach is appreciated for its capacity to generate nanofibers with diameters varying from a few nanometres to micrometres, contingent upon the characteristics of the polymer solution and the circumstances of processing. Electrospun nanofibers have a great surface area-to-volume ratio that makes them useful for a wide range of applications, including composite materials, drug delivery systems, tissue engineering scaffolds, and filtration [51].

1.5.2 Characterization Techniques

1.5.2.1 Scanning Electron Microscopy (SEM)

SEM is a potent microscopy method for high-resolution surface morphology and material structure visualization. SEM scans the surface of the object using a concentrated electron beam, as opposed to optical microscopes. Images are created by detecting signals, such as distinctive X-rays, backscattered electrons, and secondary electrons that are produced by the interaction between the electrons and atoms in the sample. Depending on the apparatus, SEM enables for the comprehensive study of materials at magnifications of up to 1,000,000×. It offers information on the elemental makeup, topography, and other surface characteristics of materials. Many disciplines, including materials science, nanotechnology, biology, geology, and forensics, make significant use of SEM [52].

1.5.2.2 Transmission Electron Microscopy (TEM)

TEM is a sophisticated microscopy method that's used to image the interior properties of materials at very high resolution. Unlike scanning electron microscopy, which scans the surface of a sample, TEM transmits a beam of electrons through a thin specimen. The electrons interact with the specimen, producing an image based on how they pass through or scatter within the sample. TEM allows for imaging at nanometre-scale resolution, revealing details of atomic structure, crystallography, and defects within materials. Depending on the instrument, TEM can be used at magnifications of up to 2,000,000×. Moreover, elemental analysis using electron energy loss spectroscopy (EELS) is made possible by TEM [53].

1.5.2.3 X-Ray Diffraction (XRD)

A potent method for examining a material's crystallographic structure is X-ray diffraction. It operates by subjecting a sample to X-rays, which then interact with the sample atom's electron clouds. The atomic arrangement inside the material causes the X-rays to diffract at precise angles, creating a diffraction pattern that can be examined to ascertain the sample's orientation, phase composition, and crystalline structure. The electrons in the atoms scatter the X-rays in various ways when they hit the sample. The atomic spacing and arrangement in the crystal lattice determine the angle and intensity of the scattered X-rays. X-ray diffraction (XRD) yields information about the crystal structure, including lattice parameters, unit cell dimensions, and crystallite size, by measuring these angles and intensities [54].

1.5.2.4 Raman Spectroscopy

Vibrational, rotational, and other low-frequency modes in a system can be studied using Raman spectroscopy. It uses monochromatic light that is inelastically scattered typically from a laser source to examine the vibrational modes of molecules. An energy shift that is unique to the vibrational modes of the molecules in the sample is the outcome of the Raman scattering. This change reveals details about the material's chemical bonding and molecular structure. When laser photons interact with the sample, the majority of them experience elastic scattering, also known as Rayleigh scattering, while a tiny percentage experience inelastic scattering, also known as Raman scattering. The vibrational energy of the chemical bonds in the sample is represented by the energy difference between incoming and scattered photons. Raman spectroscopy is used to analyse biological tissues and cells, identify chemical compositions and structures, find functional groups and impurities, study medication formulations and interactions, and guarantee uniformity and purity in manufacturing processes [55].

1.6 Applications of Nanomaterials in Specific Energy Storage Devices

1.6.1 Lithium-Ion Batteries

The high energy density and extended cycle life of lithium-ion batteries (LIBs) make them ideal for use in energy storage, electric cars, and portable devices. The performance of LIBs is greatly improved by nanomaterials, especially in the electrode materials.

1.6.1.1 Anode Materials

The anode in lithium-ion batteries (LIBs) is essential for the storage and release of lithium ions throughout cycles of charge and discharge. Compared to traditional materials, nanomaterials have several benefits because of their increased surface area, higher electrical conductivity, and capacity to withstand significant volume changes during cycling. The anode in lithium-ion batteries (LIBs) is essential for the storage and release of lithium ions throughout cycles of charge and discharge. Compared to traditional materials, nanomaterials have many benefits because of their increased surface area, higher electrical conductivity, and capacity to withstand significant volume changes during cycling. The high specific capacity of silicon (Si) nanoparticles (about 4200 mAh/g), as opposed to graphite (372 mAh/g), has attracted interest in them as a potential anode material in lithium-ion batteries [56].

However, during the insertion and extraction of lithium, silicon experiences large volume changes that result in pulverization and capacity loss. By cutting down on the tension and diffusion route length during cycle, nanosizing silicon particles helps to alleviate these problems. Graphene is a two-dimensional carbon allotrope with remarkable mechanical and electrical properties. Graphene-based materials can enhance the anode's cycling stability and rate capabilities in LIBs. To improve overall performance, graphene can be used as a conductive addition or as a scaffold for other active materials. Higher energy densities are made possible by silicon nanoparticles and graphene-based anodes, which may extend the life and capacity of LIBs. Fast charge and discharge rates are made possible by nanoparticles' improved conductivity and fewer diffusion routes. Nanomaterials can prolong the life of lithium-ion batteries and reduce electrode deterioration [57].

1.6.1.2 Cathode Materials

LIB's energy storage devices depend on cathode materials. Lithium iron phosphate ($LiFePO_4$) is a material that is frequently used in energy storage systems and electric vehicles (EVs) due to its stability and safety. It provides a long cycle life, high specific capacity, and great thermal stability. The effective insertion and extraction of lithium ions during charge and discharge cycles is made possible by the material's crystal structure, which enhances the battery's overall performance and safety. Nickel cobalt manganese oxide (NCM) cathode materials are usually composed of different proportions of manganese, cobalt, and nickel. These materials are highly regarded for their stability and high energy density. NCM cathodes are appropriate for applications needing high performance and energy efficiency, such as grid storage solutions and electric vehicles, since they strike a balance between energy density, power output, and affordability [58].

1.6.2 Supercapacitors

Supercapacitors, also known as ultracapacitors or electrochemical capacitors, are devices that store electrical energy by separating charges at the electrode-electrolyte interface. The energy density, power density, and cycle stability of supercapacitors are all significantly influenced by the materials used for the electrodes. Carbon-based materials, including activated carbon and carbon nanotubes (CNTs), are extensively used as electrode materials in supercapacitors because of their large surface area, electrical conductivity, and electrochemical stability. For instance, CNTs have superior electrical and mechanical characteristics, which makes them a good fit for high-performance supercapacitors. Additional materials used to improve the capacitance and energy density of supercapacitors are metal oxides and conducting polymers [59].

Metal oxides, such as manganese dioxide (MnO_2), ruthenium oxide (RuO_2), and nickel oxide (NiO), offer better capacitance than carbon-based materials; however, they may have limits in terms of cycle life and cost. Conducting polymers like polyaniline (PANI) and polypyrrole (PPy) are appealing due to their reversible redox processes, which contribute to capacitance. The goal of research in India and throughout the world is still to create electrodes based on nanomaterials that have better electrochemical performance and stability for use in supercapacitors. The commercialization of supercapacitors with greater energy density, longer cycle life, and cheaper prices is anticipated as a result of advances in nanotechnology, which will facilitate their incorporation into a range of energy storage devices [60].

Electrolytes serve an important role in supercapacitors by promoting ion movement between electrodes, which is required for electrical energy storage and release. The energy density, power density, and stability of supercapacitors are among the performance characteristics that are highly impacted by the choice of electrolyte. Supercapacitorsfrequently employ aqueous electrolytes, which are usually based on aqueous solutions of salts like potassium hydroxide (KOH), sodium sulphate (Na_2SO_4), or sulphuric acid (H_2SO_4). These electrolytes have a high ionic conductivity, are reasonably priced, and are eco-friendly. However, the water's electrochemical stability window limits its operational voltage [61].

Organic electrolytes are used in non-aqueous supercapacitors, and organic electrolytes typically consist of organic solvents that include acetonitrile (ACN), propylene carbonate (PC), or tetraethylammonium tetrafluoroborate (TEABF4). Compared to aqueous electrolytes, they offer a larger electrochemical stability window that permits greater working voltages and energy densities. However, because they are poisonous and flammable, organic electrolytes could be dangerous. Another family of electrolytes with strong thermal stability, minimal volatility, and broad electrochemical stability are ionic liquids. To increase energy density and safety in supercapacitors, they have been studied for usage. To address the problems with leakage, evaporation, and safety associated with liquid electrolytes, solid-state electrolytes, such as polymer and ceramic electrolytes, are now being investigated. These electrolytes can raise voltage operation, improve device stability, and maybe boost energy density. Research in India and throughout the world is concentrated on creating sophisticated electrolytes that can enhance the efficiency and dependability of supercapacitors, enabling them to be used in a variety of applications, which include energy storage devices, hybrid cars, and portable gadgets [62].

1.6.3 *Fuel Cells*

Fuel cells are electrochemical devices that produce electrical energy directly from the chemical energy contained in fuels. In order to facilitate fuel cell reactions, increase their efficiency, and lower the overpotential needed for reactions to occur, catalysts are essential. The most popular catalysts used in proton exchange membrane fuel cells (PEMFCs) are platinum (Pt) nanoparticles. When it comes to the oxygen reduction reaction (ORR) at the cathode and the hydrogen oxidation reaction (HOR) at the anode, Pt catalysts show exceptional catalytic activity and endurance. Pt may, however, be costly and subject to dissolution and carbon monoxide (CO) poisoning over time, leading to its eventual disintegration [63]. Research focuses on creating non-precious metal catalysts (NPMCs) to overcome the availability and cost concerns related to Pt. Nitrogen-doped carbon materials, carbon-based nanomaterials like graphene and carbon nanotubes, and transition metal-based catalysts like Fe, Co, and Ni have demonstrated encouraging catalytic activity and stability for fuel cell processes. The goal of these catalysts is to preserve or even increase overall performance while lowering reliance on pricey metals like Pt. In addition to the active catalyst, the support material affects the efficiency of fuel cells. Carbon-based materials with high surface area, electrical conductivity, and stability in fuel cell environments, including carbon black and carbon nanotubes, are frequently utilized as supports [64].

1.7 Challenges and Future Directions

1.7.1 *Scalability of Nanomaterial Production*

Realizing the broad application of nanomaterials in a variety of industries, such as energy storage, electronics, and health care, is still complicated by the scalability of nanomaterial manufacturing. The efficiency and cost-effectiveness with which several nanomaterial synthesis techniques, including hydrothermal synthesis, sol-gel, and chemical vapour deposition (CVD), can create vast amounts of nanomaterials is frequently restricted. It is a challenging issue to scale up these procedures while preserving the desirable characteristics and homogeneity of nanomaterials. Commercialization depends on large-scale nanomaterial manufacturing being cost-effective. To make nanomaterials competitive with traditional materials, high costs related to equipment, energy inputs, and raw materials must be reduced in the synthesis processes [65].

Sustainable practices in the manufacture of nanomaterials can raise environmental concerns. Green synthesis techniques that reduce their negative effects on the environment are necessary because certain synthesis procedures may use dangerous chemicals or produce undesirable by-products. It is difficult to guarantee the repeatability and constant quality of nanomaterials at large scale. Changes in the characteristics of particles, such as size, shape, purity, and surface chemistry, can impact their output and result in applications. There is potential for the development of innovative synthesis techniques including electrospinning and template-assisted growth that are scalable, economical, and environmentally safe; optimization of already used manufacturing methods to increase material characteristics, lower energy use, and increase yield; and incorporation of nanoparticles for use in biological sensors and energy storage devices into scalable production processes. To guarantee the safe integration of nanomaterials into production and usage processes, regulatory frameworks and safety criteria must be established [66].

1.7.2 Integration into Existing Technologies

Addressing concerns about nanomaterial-integrated devices long-term durability and dependability is critical. For practical applications, it is crucial to guarantee that nanomaterials retain their characteristics in the face of changing operating stressors and environmental conditions. Finding economical ways to produce nanomaterials and incorporate them into current technology is still a major issue. Widespread adoption requires the cost-effective scaling up of production processes while upholding quality and performance criteria. It is essential to comprehend the possible health and safety hazards connected to nanoparticles at the stages of manufacture, integration, and end-use. It is crucial to create guidelines and standards to reduce these dangers and guarantee secure handling and disposal procedures. To maximize the performance of nanomaterial-integrated technologies, ongoing efforts are required. This entails increasing lifespans, lowering energy usage, boosting conductivity, and improving efficiency.

It is crucial to get beyond obstacles to market uptake, such competition from traditional technologies, customer acceptability, and regulatory clearances. Accelerating commercialization requires proving the concrete advantages and competitive advantages of integrating nanomaterials; assisting with current research and development initiatives to investigate new nanomaterials, cutting-edge integration techniques, and cutting-edge applications in a variety of fields, such as electronics, energy storage, medical, and environmental remediation; and bringing together worldwide norms and laws pertaining to the integration of nanomaterials to enable commerce internationally, guarantee product safety, and foster innovation under a unified regulatory framework [67].

1.7.3 Environmental and Health Concerns

Nanomaterials provide promising advances in a variety of sectors, but their extensive usage creates environmental and health issues. When discharged into the environment, nanoparticles' tiny size and special qualities may have unforeseen effects. Their possible bioaccumulation in ecosystems, interactions with other creatures, and persistence in natural systems are all causes for concern.

Exposure at the stages of manufacture, use, and disposal poses health concerns. Through their ability to cross biological boundaries and interact with cells, nanoparticles may have harmful consequences, such as oxidative stress, inflammation, or other impacts. Moreover, the absence of all-encompassing regulatory structures contributes to the uncertainty surrounding their secure management and elimination. Safe production, usage, and environmental discharge criteria also need to be developed [68].

1.7.4 Future Trends

The remarkable features of advanced materials, especially two-dimensional materials, such as graphene and transition metal dichalcogenides (TMDs), have the potential to completely transform a range of sectors. Graphene has excellent mechanical strength, electrical conductivity, and flexibility, which makes it perfect for use in composite materials, electronics, and energy storage. Transition metal dichalcogenides have semiconducting characteristics that are essential for optoelectronics and next-generation electronics. Their atomically thin structure makes it possible to precisely manipulate the characteristics of the materials, leading to advances in photovoltaics, catalysis, and sensors. In future, research will concentrate on improving the scalability and manufacturability of 2D materials, incorporating them into current technologies, and investigating new uses in environmental remediation and medicinal devices [69].

Novel strategies for the synthesis of nanomaterials are essential for developing material science and increasing application possibilities. Techniques like atomic layer deposition (ALD), which allows for exact atomic-level control in thin film formation and is appropriate for electrical and catalytic applications, have been the focus of recent trends. Hydrothermal synthesis provides an environmentally benign and economically viable way to generate nanomaterials under moderate circumstances, whereas molecular beam epitaxy (MBE) is essential for creating complex nanostructures with exact composition and crystalline characteristics [70].

Hybrid nanomaterials integrate different constituents, capitalizing on their complementary attributes to optimize efficacy across a range of uses. To improve electrical conductivity and catalytic activity, carbon nanotubes and metal nanoparticles can be combined. These properties are important for energy storage and environmental clean-up. Integrating polymers with quantum dots enhances optical qualities and biocompatibility for sophisticated imaging and medication delivery systems in biomedical applications. Subsequent investigations centre on enhancing interfaces in hybrid systems to attain exceptional features customized for distinct industrial and biological requirements [71, 72].

1.8 Conclusion

Nanomaterial's high energy density, long cycle life, and quick charge/discharge rates have completely changed the energy storage industry. The energy conversion of fuel cells, batteries, and supercapacitors is optimized by their nanoscale properties, such as increased surface area and higher catalytic activity. Different kinds metal oxides, sulphides, carbon-based, and nanocomposites are customized using methods including hydrothermal synthesis and sol-gel. Their structures are clarified by characterization using Raman spectroscopy, XRD, TEM, and SEM. Environmental effects, integration, and scalability are among the difficulties. Future developments will centre on cutting-edge materials, such as hybrid nanomaterials and 2D structures, which hold promise for cooperative research towards sustainable energy solutions.

Nanomaterials have tremendous potential to alter future energy storage systems by increasing efficiency, capacity, and sustainability. The energy density, charge/discharge rates, and cycle life of batteries, supercapacitors, and fuel cells may all be significantly increased because of their special nanoscale characteristics, which include large surface area and customizable electronic architectures. As research on environmental compatibility, cost-effectiveness, and scalability continues, nanomaterials are well positioned to meet the world's energy needs while lowering dependency on fossil fuels. A new era of cleaner, more effective energy solutions for a sustainable future is promised by their incorporation into common energy storage applications.

Innovations in synthesis processes and nanomaterials have propelled a remarkable development in energy storage technology. Improvements in energy density, cycle life, and efficiency have been made in a number of applications, including fuel cells, batteries, and supercapacitors. Significant advancements have been made possible by the nanoscale characteristics of materials like metal oxides and carbon-based composites. Although problems with integration and scalability still exist, further research on advanced materials indicates that development will continue. Future developments might bring forth new manufacturing techniques and hybrid systems, among other advancements, which would provide effective and sustainable energy storage solutions for a range of technical requirements.

References

1. Zhang Y et al (2016) Nanomaterials: properties, applications, and environmental impact. J Nanotechnol 12(3):123–145
2. Khan Y et al (2019) Fundamentals of nanomaterials: classification and applications. Adv Mater 31(45):1902567
3. Liu Y et al. (2018) Nanomaterials in drug delivery: challenges and perspectives. Nanomed Nanotechnol Biol Med 14(6):1607–1624
4. Chen X et al (2020) Nanomaterials: a review of their characteristics and applications. Mater Sci Eng 456:23–39
5. Lee J et al (2017) Nanomaterials for environmental remediation: a review. Environ Sci Technol 51(8):4642–4660
6. Wang J et al (2021) Nanomaterials in energy storage systems: review and perspectives. Energy Storage Mater 34:168–179
7. Nair S et al (2020) Nanomaterials in medicine: a review. Biotechnol Adv 38:107341
8. Liu J et al (2020) Energy storage technologies and applications in smart grids. Renew Sustain Energy Rev 120:109674
9. Zhao Y et al (2019) Energy storage systems for grid stabilization and frequency regulation. Electr Power Syst Res 171:54–65
10. Khan MI et al (2020) Economic analysis of energy storage systems: a case study. Energy Econ 85:104548
11. Hodge B-M, Ault G (2021) Environmental benefits of energy storage systems: a review. Environ Sci Technol 55(18):12234–12247
12. Nykvist B, Nilsson M (2015) Rapidly falling costs of battery packs for electric vehicles. Nat Clim Chang 5(4):329–332
13. Zhu Y et al (2018) Nanomaterials for energy storage applications. Chem Rev 118(1):1–10
14. Liu H et al (2020) Advanced nanomaterials for supercapacitor applications. J Power Sour 451:227768
15. Kumar S et al (2021) Nanotechnology for water purification: a review. J Environ Manage 277:111407
16. Rana F et al (2020) Nanomaterials in electronics: applications and advances. Nat Electron 3(7):501–516
17. Zhang W et al (2019) Targeted nanomedicine for cancer therapy. Front Pharmacol 10:100
18. Sarkar S et al (2020) Nanomaterials for renewable energy applications. Renew Sustain Energy Rev 120:109674
19. Khan Y et al (2021) Nanomaterials for energy efficiency and sustainability. Mater Today 45:12–21
20. Goodenough JB, Park KS (2013) The Li-ion battery: a perspective. J Am Chem Soc 135(4):1167–1176
21. Scrosati B, Garche J (2010) Lithium batteries: status, prospects and future. J Power Sour 195(9):2419–2430
22. Li X et al (2020) Recent advances in solid-state lithium-ion batteries. Energy Storage Mater 32:54–82
23. Wang Y, Zhang H, Wang J (2018) Lead-acid battery technology and environmental management. J Power Sour 400:184–191
24. Chen H, Zhang Y, Liu J (2017) Nickel-metal hydride batteries: performance and applications. Energy Rev Res 56:150–162
25. Miller JR, Simon P (2008) Electrochemical capacitors for energy management. Science 321(5889):651–652
26. Zhao Y, Wang D (2016) High-performance supercapacitors with 3D nanostructured electrodes. Nano Energy 30:327–343
27. Rao ENM et al (2020) Recent trends in high-temperature fuel cells: an overview. Fuel Cells 20(5):675–692

28. Schuster JWH et al (2019) Techno-economic analysis of fuel cell systems for stationary and mobile applications. Energy Convers Manage 183:1182–1194
29. Wang HT et al (2021) A review on solid-state battery technology and its advancements. J Power Sour 478:229183
30. Zhang LM et al (2020) Solid-state batteries: challenges and opportunities. Energy Environ Sci 13:6456–6483
31. Narayan SV et al (2020) A review on flow batteries: recent developments and applications. J Energy Storage 32:101811
32. Sharma RK et al (2021) Progress and challenges in flow battery technology for energy storage applications. Renew Sustain Energy Rev 135:110191
33. Patil AG et al (2021) Advancements in flow battery technology for renewable energy storage. Mater Today Proc 42:445–451
34. Agrawal SK et al (2021) Nanostructured materials for enhanced charge/discharge rates in energy storage systems. J Nanosci Nanotechnol 21(10):5484–5493
35. Patel AS et al (2020) Nanotechnology for rapid charge/discharge rates in lithium-ion batteries. J Power Sources 484:229280
36. Kumar S et al (2021) Nanomaterials for enhanced cycle life in lithium-ion batteries. J Nanosci Nanotechnol 21(11):6332–6341
37. Patel A et al (2021) Improving cycle life with nanostructured electrodes in energy storage systems. J Energy Storage 36:102473
38. Verma AK et al (2021) Nanomaterial innovations for cost reduction in battery technologies. Mater Sci Eng B 270:115222
39. Rao MS et al (2021) Affordable energy storage through nanotechnology. Energy Storage Mater 34:222–231
40. Reddy SA et al (2021) Graphene-based materials for energy storage: a review. J Nanosci Nanotechnol 21(8):4137–4151
41. Patel PR et al (2020) Carbon nanotube composites for high-performance supercapacitors. Adv Func Mater 30(35):2002726
42. Yadav AK et al (2021) Enhanced electrochemical performance of supercapacitors using fullerenes. Nano Energy 79:105383
43. Desai LS et al (2020) Fullerenes as high-performance electrode materials for batteries. Adv Func Mater 30(44):2003006
44. Verma SK et al (2020) Transition metal oxides in energy storage devices: a review. J Energy Storage 32:101823
45. Kumar A et al (2021) Molybdenum disulfide (MoS_2) in energy storage applications: a review. Mater Today Chem 20:100491
46. Sharma S, Kumar M, Yadav R (2020) Metal-organic frameworks (MOFs) in energy storage systems. J Energy Storage 32:101–112
47. Patel R, Singh N (2020) Advanced metal sulfide nanocomposites for high-performance supercapacitors. J Nanosci Nanotechnol Res 15(3):134–148
48. Hachem K et al (2022) Methods of chemical synthesis in the synthesis of nanomaterial and nanoparticles by the chemical deposition method: a review. BioNanoScience 12(3):1032–1057
49. Shah K et al (2019) Recent advances in sol-gel process for nanomaterials and their applications. Mater Today Proc 16:646–652
50. Zhao X et al (2021) Hydrothermal synthesis of metal oxide nanoparticles: mechanisms, methods, and applications. Chem Rev 121(1):403–426
51. Reneker DH, Yarin AL (2008) Electrospinning jets and polymer nanofibers. Polymer 49(10):2387–2425
52. Kumar M (2008) Applications of scanning electron microscopy in pharmaceutical research. J Pharmacy Bioallied Sci 1(2):97–100
53. Egerton RF (2011) Physical principles of electron microscopy: an introduction to TEM, SEM, and AEM. Springer Science & Business Media
54. Cullity BD, Stock SR (2001) Elements of X-ray diffraction. Prentice Hall
55. Ekins S (2007) Pharmaceutical applications of Raman spectroscopy. Wiley

56. Wang Q (2021) Silicon-based composites as anodes for lithium ion batteries
57. Lavagna L et al (2020) Graphene and lithium-based battery electrodes: a review of recent literature. Energies 13(18):4867
58. Konarov A, Myung S-T, Sun Y-K (2017) Cathode materials for future electric vehicles and energy storage systems. ACS Energy Lett 2(3):703–708
59. Zhong C et al (2015) A review of electrolyte materials and compositions for electrochemical supercapacitors. Chem Soc Rev 44(21):7484–7539
60. Jayakumar S et al (2023) A comprehensive review of metal oxides (RuO_2, Co_3O_4, MnO_2 and NiO) for supercapacitor applications and global market trends. J Alloys Compounds: 173170
61. Majumdar D (2021) Aqueous electrolytes for flexible supercapacitors. In: Flexible supercapacitor nanoarchitectonics, 349–412
62. Truong TTT et al (2024) Conventional supercapacitor electrolytes: aqueous, organic, and ionic. In: Supercapacitors. Elsevier, 245–265
63. Tellez-Cruz MM et al (2021) Proton exchange membrane fuel cells (PEMFCs): advances and challenges. Polymers 13(18):3064
64. Zhou X et al (2014) A review of graphene-based nanostructural materials for both catalyst supports and metal-free catalysts in PEM fuel cell oxygen reduction reactions. Adv Energy Mater 4(8):1301523
65. Kokila GN, Mallikarjunaswamy C, Lakshmi Ranganatha V (2022) A review on synthesis and applications of versatile nanomaterials. Inorganic Nano-Metal Chem: 1–30
66. García-Quintero A, Palencia M (2021) A critical analysis of environmental sustainability metrics applied to green synthesis of nanomaterials and the assessment of environmental risks associated with the nanotechnology. Sci Total Environ 793:148524
67. Epelle EI et al (2022) A comprehensive review of hydrogen production and storage: a focus on the role of nanomaterials. Int J Hydrogen Energy 47(47):20398–20431
68. Ahmed SF et al (2022) Green approaches in synthesising nanomaterials for environmental nanobioremediation: technological advancements, applications, benefits and challenges. Environ Res 204:111967
69. Choi W et al (2017) Recent development of two-dimensional transition metal dichalcogenides and their applications. Mater Today 20(3):116–130
70. Zhao Y et al (2021) Atomic/molecular layer deposition for energy storage and conversion. Chem Soc Rev 50(6):3889–3956
71. Rao R et al (2018) Carbon nanotubes and related nanomaterials: critical advances and challenges for synthesis toward mainstream commercial applications. ACS Nano 12(12):11756–11784
72. Saleh HM, Hassan AI (2023) Synthesis and characterization of nanomaterials for application in cost-effective electrochemical devices. Sustainability 15(14):10891

Chapter 2
Fundamentals of Computational Design in Nanomaterials

S. Sreejith, J. Ajayan, N. V. Uma Reddy, J. M. Radhika, Jimsha K. Mathew, B. Sivasankari, and N. V. Raghavendra Reddy

Abstract Nanomaterials (NMs) have drawn a lot of interest in materials science and engineering due to their unique properties and diverse range of applications. Computational design (CD) approaches are crucial for understanding and optimizing the properties of NMs, as they offer valuable insights that improve experimental procedures. An overview of the core ideas and techniques used in CD for NMs is provided in this chapter. Some of the key computational approaches we discuss are Monte Carlo (MonC) simulations, density functional theory (DenFT), coarse-grained modelling (CGM), molecular dynamics simulations (MolDS), and machine learning (MAL) techniques. These methods can help researchers focus their experimental synthesis and characterization efforts, predict how NMs would behave in certain scenarios, and look into the mechanical, electrical, and structural properties of NMs. The field's future goals and challenges are also discussed, with an emphasis on how interdisciplinary collaboration and on-going innovation are required to address difficult NMs design problems. This study is a significant resource for researchers and practitioners interested in learning more about the exciting topic of CD in NMs since it provides insights into the methodology, applications, and potential effect of this field.

Keywords Molecular dynamics simulations · Machine learning · Computational design · Nanomaterials · Monte Carlo simulations

---

S. Sreejith (✉) · N. V. Uma Reddy · J. K. Mathew
New Horizon College of Engineering, Bengaluru, Karnataka, India
e-mail: dr.sreejiths.nhce@newhorizonindia.edu

J. Ajayan · J. M. Radhika
SR University, Warangal, Telangana, India

B. Sivasankari
SNS College of Technology, Coimbatore, Tamil Nadu, India

N. V. Raghavendra Reddy
RV Institute of Technology and Management, Bengaluru, Karnataka, India

R. Naik et al. (eds.), *Advanced Nanomaterials for Energy Storage Devices*, Nanostructure Science and Technology, https://doi.org/10.1007/978-3-031-74730-4_2

2.1 Introduction

A comprehensive exploration of the nanoscale domain has been motivated by materials science's quest for novel materials with tailored properties. NMs differ from their bulk counterparts in mechanical, chemical, and physical properties due to their nanoscale size. Using these unique qualities has the potential to transform a number of industries, including electronics, energy storage, medicine, and environmental clean-up. However, the design and synthesis of NMs present formidable challenges due to the complexity of nanoscale phenomena and the intricate interplay of parameters governing their behaviour. In response, CD has emerged as a powerful tool, offering insights into the underlying principles governing NMs properties and enabling the rational design of materials with tailored functionalities [1–5]

Various computational approaches play a crucial role in designing and understanding NMs (Fig. 2.1). MolDS track the movements of atoms and molecules, offering insights into NMs' dynamic behaviour and properties. DenFT utilizes quantum mechanics to calculate electronic structures, aiding predictions of properties like band gaps and charge transfer. MonC simulations simulate NMs behaviour based on probability distributions, especially useful for studying thermodynamics and phase transitions. CGM simplifies descriptions by grouping atoms, enabling study of larger scales and phenomena like self-assembly. Although they provide precise descriptions, Ab Initio (AbIn) methods are computationally costly. Large datasets are used by MAL to create predictive models, which speeds up the discovery of new materials. Time evolution is modelled via kinetic MonC simulations, which are helpful for growth mechanisms and surface responses. Simulations combining quantum physics and classical descriptions, such as molecular mechanics or quantum physics, are useful for comprehending interactions with intricate settings. Together, these varied approaches improve our knowledge and design of NMs, spurring innovation across a range of industries [6–15]. In this chapter, the foundations of CD in NMs are explored, along with the approaches, theories, and uses that support this multidisciplinary subject. This chapter has synthesized theoretical frameworks, computational techniques, and experimental validations to study the subject of NM design.

2.2 Computational and Theoretical Foundations

In order to improve our comprehension and capacity for design at the atomic and molecular dimensions, the computational and theoretical underpinnings of NM research are essential. The majority of this research is based on quantum mechanics, which makes use of methods like DenFT to predict structural and electrical properties. By solving the Schrödinger equation (Eq. 2.1) for complex many-electron systems, DenFT yields precise insights into electronic densities and makes accurate predictions about the behaviour and features of materials. [16].

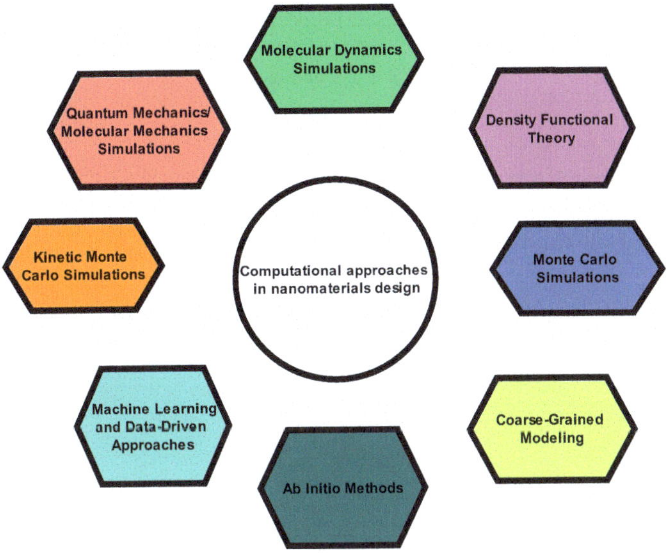

Fig. 2.1 Computational approaches in NMs design

$$\hat{H}\Psi(x_1, x_2, \ldots, x_N) = E\Psi(x_1, x_2, \ldots, x_N) \tag{2.1}$$

where E is the total energy of the system, Ψ is the many-electron wave-function, and $\hat{H}$ represents the Hamiltonian operator.

MolDS solve Newton's equations of motion (Eq. 2.2) for a system of atoms in addition to DenFT, offering a dynamic view of atomic interactions across time. This approach is particularly helpful for studying phenomena including phase transitions, diffusion, and defect dynamics, as well as mechanical characteristics like stress and strain responses [17].

$$m_i \frac{d^2 r_i}{dt^2} = F_i \tag{2.2}$$

where r_i represents its position, F_i is the force acting on it, and m_i indicates the mass of atom i.

Moreover, phase behaviour and thermodynamic properties are studied using MonC simulations by statistical sampling, and the mechanical behaviour of materials with complex geometries is represented using finite element analysis. The content is divided into finite elements linked at nodes in finite element analysis using the field variable (Eq. 2.3). To estimate properties, MonC methods require random sampling (Eq. 2.4). [18]

$$u(r) \approx \sum_{i=1}^{N} N_i(r) u_i \tag{2.3}$$

where u_i represent nodal values and $N_i(r)$ represents shape function.

$$A = \int A(r)P(r)\mathrm{d}r \tag{2.4}$$

where $P(r)$ represents probability density function.

MAL approaches are utilized to enhance computational toolkit by facilitating rapid material property prediction and material identification through analysis of large datasets from experiments and simulations. When coupled, these computational methods provide a solid basis for the optimization and innovative design of NMs. The following mathematical ideas are commonly used to predict material properties and identify new materials using MAL. Using feature vectors to capture their properties, materials are represented (Eq. 2.5) [19].

$$x = [x_1, x_2 \ldots x_n] \tag{2.5}$$

where x represents individual features.

The relationship between the feature vector x and the target property y is modelled using a prediction function (x) (Eq. 2.6) [19].

$$y = f(x) + \varepsilon \tag{2.6}$$

where $f(x)$ represents the model and ε represents the error term.

NM research is revolutionized by combining computational techniques like MAL, MolDS, DenFT, finite element analysis, and MonC simulations. These methods enable enhanced material design through accurate predictions, dynamic insights, and rapid discovery. Technology and materials science have advanced significantly as a result of this multidisciplinary approach.

2.3 Various Applications of Computational Techniques in Nanomaterials Design: Recent Advances

As a vital tool for creating, describing, and comprehending NMs, MolDS offer atomistic insights into how they behave in various contexts. These models could provide crucial information on shape, surface properties, and defects by foretelling atomic-scale structures. By simulating assembly processes or growth, researchers can gain a better understanding of features and final structures that are influenced by factors including surface functionalization and stability effects, or the self-assembly of nanoparticles into ordered structures. Additionally, the design of NMs for application in nano-electronics, nano-composites, and biomaterials is aided by the examination of mechanical qualities, such as elasticity, strength, and deformation behaviour made possible by MolDS. For the purpose of creating materials for gas separation, molecular sieving, and drug delivery all of which profit immensely from MolDS, it is essential to comprehend diffusion mechanisms and transport properties. Moreover, thermodynamic characteristics, such as free energy, entropy, and phase equilibria, which offer details on phase transitions, stability, and chemical reactions are computed using MolDS. They study surface chemistry and interfaces, which are important for controlling surface properties, catalytic activity, and interactions with other materials or biomolecules. By offering atomistic-level insights, MolDS augment experimental methods and enable the logical design of materials suitable for a broad range of applications. Few recent application of MolDS simulation in NM design and analysis is highlighted below. In the year 2019, Atilhan et al. [20] investigated using classical MolDS the nanoscopic properties of a mixed deep eutectic solvent consisting of choline chloride as the hydrogen bond acceptor, and urea, ethylene glycol, and glycerol as the hydrogen bond donor. For several of the relevant hydrogen bond donors, the establishment of predominant hydrogen bonding suggests local heterogeneities. Different locations across the contact also determine the interfacial behaviour at vacuum. Minor structuring occurs at liquid–liquid interfaces due to the existence of a common type of hydrogen bond acceptor. When graphene and boron nitride are taken into account, mixed deep eutectic behaviour on two-dimensional surfaces reveals high adsorption and minimal layering for MoS_2, but the adsorbed layers on these materials are likewise heterogeneous. Analyses of the behaviour of carbon nanotubes have also been conducted; these have corroborated the preferential confinement of ethylene glycol, one of the suggested hydrogen bond donors, and the substantial hydrogen bond acceptor adsorption that occurs during nanotube solvation. The obtained results demonstrate heterogeneities in the mixed deep eutectic solvent at the nanoscale level caused by the created preferred hydrogen bonding, which are suggested to regulate the composition of deep eutectic solvents. In the year 2021, Abbaspour et al. [21] reported using MolDS with 350 MPa of applied pressure to study the effectiveness of nanostructures resembling carbon peapods made up of 1 and 3 fullerenes into (13,13) armchair carbon nanotubes in the desalination process. Peapods are a new class of nanostructures that have received a great deal of attention and applications in the past few years due to their unique thermal and electrical properties compared to regular carbon nanotubes. Numerous features have been computed, such as the flux of water and ions (Fig. 2.2), the self-diffusion coefficient, the radial distribution function, the potential of mean force, the radial distribution function, the analysis of hydrogen bonds, and the penetration barrier energy. According to the findings, the pure carbon nanotubes system (2565 ns^{-1}) and the fullerene peapod reverse charged system (270 ns^{-1}) had the greatest and lowest water flows, respectively. Among the other systems, the pure carbon nanotubes exhibited the highest ion flow (26 ns^{-1}) as well.

In the year 2024, Borówko et al. [22] reported the findings from extensive MolDS of solid-surface adsorption by nanoparticles. The particles were represented as rigid groups of spherically segmented particles. Three different particles types namely rectangles, triangles, rods with an equal number of segments were examined. M. Borówko et al. demonstrated the effects of particle shape on adsorption, surface layer structure, and the extent of particle removal from the solvent. Systems with various interactions between segments and between segments and surfaces as well

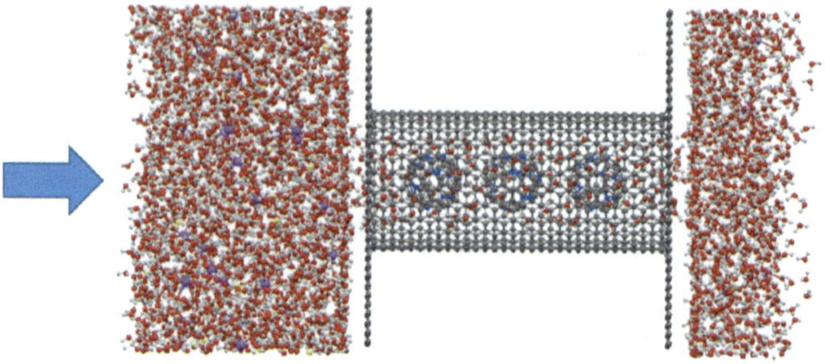

Fig. 2.2 Three N-doped fullerenes system snapshot following two nanoseconds of simulation [21] (Reused with permission from Ref. [21] Copyright 2021 Elsevier)

as varying particle concentrations were examined. The arranged configurations that emerged in adsorption monolayers were also examined. In the year 2024, Das et al. [23] studied mechanical characteristics of plumbene (PMB) by MolDS simulation study that employs equilibrium, bending, and oscillation methods. These days, two-dimensional NMs are given enormous precedence because of their unique structure and extraordinary features. The recently discovered two-dimensional allotrope of lead, called PMB, has a single-layered, hexagonal atomic structure similar to that of stanene, silicene, graphene, and germanene. A PMB sheet is created using multi-scale modelling in order to assess the mechanical characteristics of PMB under various conditions. Tensile modelling was used to study PMB's mechanical properties in the past. Using MolDS modelling, PMB sheets under various loading scenarios are investigated in order to provide an overall view of the material's mechanical properties for effective application in a variety of engineering applications. Using bending, oscillation, and equilibrium MolDS techniques, Young's modulus of PMB is calculated and compared to a value previously published from tensile modelling of the material. Evaluation is also done on how different loads affect the PMB sample in comparable circumstances. Utilizing the equilibrium MolDS method, the bulk's modulus and Poisson's ratio of PMB are also determined.

With its crucial insights into the electrical properties, chemical reactivity, and structural stability of NMs, DenFT has become an indispensable tool in NM CD. DenFT is a useful tool for rationalizing experimental data, guiding material synthesis, and identifying novel materials with desired features since it can accurately predict material properties at the quantum mechanical level. For example, simulations using DenFT have aided in our understanding of the electronic band structure and charge transport mechanisms of NMs utilized in photo catalysis, light-emitting diodes, and photovoltaics. Moreover, the development of novel catalysts for energy conversion and environmental remediation has been facilitated by the screening of candidate materials using DenFT. DenFT is still a powerful tool in the design of NMs, promoting creativity and enabling the rational development of materials with particular properties despite its high computing cost. In the year 2023, Adara et al. [24] reported simulation and modelling of $Al_2O_3^-$ CNT(3,3) CAO NMs mechanical and electrical characteristics using DenFT. Software from VASP was used to implement the spin-polarized DenFT computations using the Vienna AbIn Simulation Package (Fig. 2.3). Based on an all-electron technique known as the Perdew–Burke–Ernzerhof exchange–correlation functional, the calculations were performed using the generalized-gradient approximation. In comparison with Al_2O_3, which is primarily utilized as an insulator in microelectronics, the NM displays an exceptional energy gap of 0.000 eV and 0.007 eV, or about equivalent to conductors for downhill and upward spin, respectively.

In the year 2023, Bassey et al. [25] examined the co-doping of B, N, P, and S with Ag-doped silicene in order to assess their potential as adsorbents for the delivery of dacarbazine. Their sensitivity, conductivity, stability, and reactivity were investigated at the theory level of ωB97XD/def2SVP utilizing DenFT computations (Fig. 2.4). The findings of the geometry optimization showed that the Ag52–Si30 bond in the Ag-functionalized silicene nano surface was greatly lowered by the addition of B, N, P, and S as co-dopants, going from 2.589 Å to a range of 2.241–2.074 Å. Similarly, the energy gaps showed a similar post-co-doping magnitude reduction impact, with the interactions falling between 3.1186 and 3.7325 eV. With respect to adsorption properties, the Ead values, which were 28.399, 147.445, 235.100, and − 141.345 kcal/mol, respectively, showed physisorption in the B, N, and P-co-doped contacts and chemisorption in the S-co-doped system. After applying

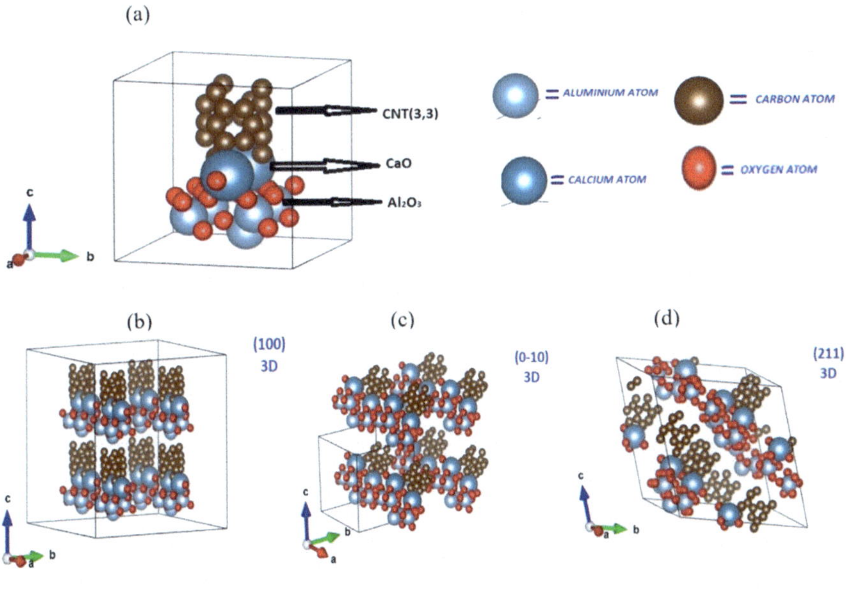

Fig. 2.3 **a** NM Al_2O_3–CaO-CNT (3,3) created using VESTA software with orientation. **b** 3-dimension (100). **c** 3-dimension (0–10). **d** 3-dimension (211) [24] (Reused with permission from Ref. [24] Copyright 2023 Elsevier)

the basis set superposition error correction to the computed adsorption energies, the modified results were as follows: dcb_B@AgSi exhibited 28.400 kcal/mol, dcb_P@AgSi displayed 135.103 kcal/mol, and dcb_N@AgSi showed 147.446 kcal/mol. In the meantime, the adsorption energy of dcb_S@AgSi was − 142.344 kcal/mol. In addition, partial and covalent contacts as well as non-covalent interactions were found when QTAIM and NCI were used to analyse the data.

In the year 2024, Adara et al. [26] reported using DenFT to examine the mechanical and electrical characteristics of Al_2O_3–CaO-CNT (3,3) NM in the presence of the VAl flaw, which is a defect in Al_2O_3 that has a lower formation energy. Al_2O_3–CaO-CNT(3,3) NM is proven to have outstanding mechanical and electrical properties without any flaws, making it appropriate for use in fuel cells, batteries, and microelectronics (Fig. 2.5). Using an all-electron approach and the generalized-gradient approximation, the Perdew–Burke–Ernzerhof exchange–correlation functional was used in the calculations. The defect caused the NMs spin-up and spin-down energy gaps to rise, from 0.007 eV to 0.000 eV, or 0.560 eV and 0.129 eV, respectively. These

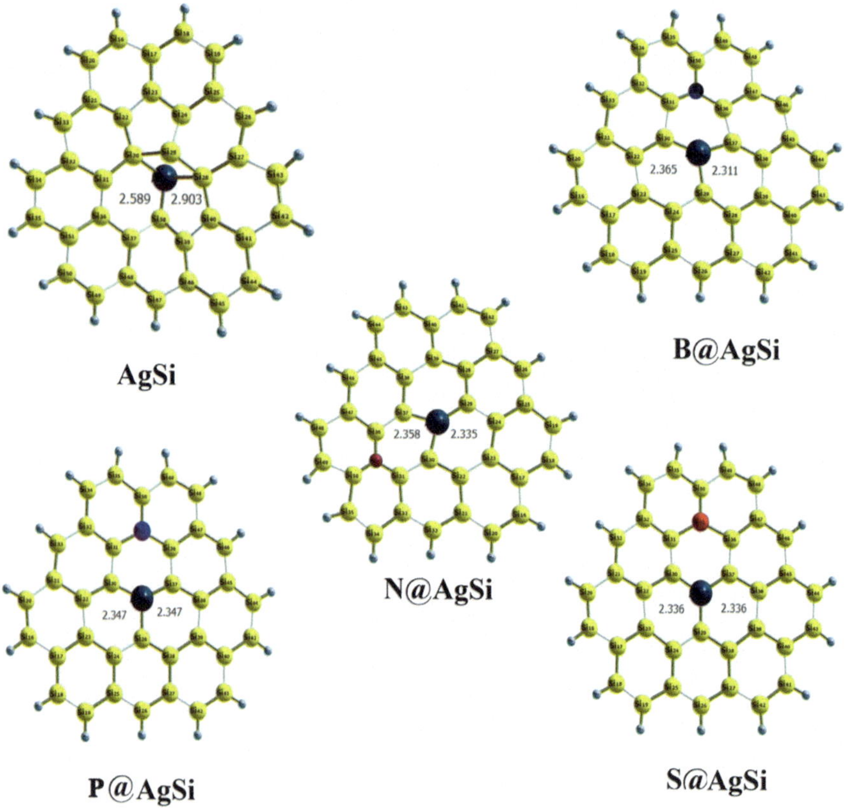

Fig. 2.4 Optimized structures of nano-surfaces at ωB97XD/def2SVP level of theory [25] (Reused with permission from Ref. [25] Copyright 2023 Elsevier)

numbers are still less than those of Al_2O_3, an insulator that is frequently utilized. For spin down and spin up, respectively, the lowest and greatest energy gap values were measured at 0 °C and 60 °C, and at 0 °C and 80 °C. Some of the mechanical properties of the NM also deteriorated as a result of the defect, primarily along the 0–10 range, where the compressibility increased from 56.82 to 90.68 GPa and the highest bulk modulus dropped to 8.40 from 48.73 GPa.

In the year 2021, Mirzaei et al. [27] reported using Monte Carlo simulation for design and optimization of a luminescent solar concentrator based on graphene quantum dots. The global energy crisis and climate change are driving up demand for green energy these days. If the cost of the photovoltaic cell can be reduced, solar

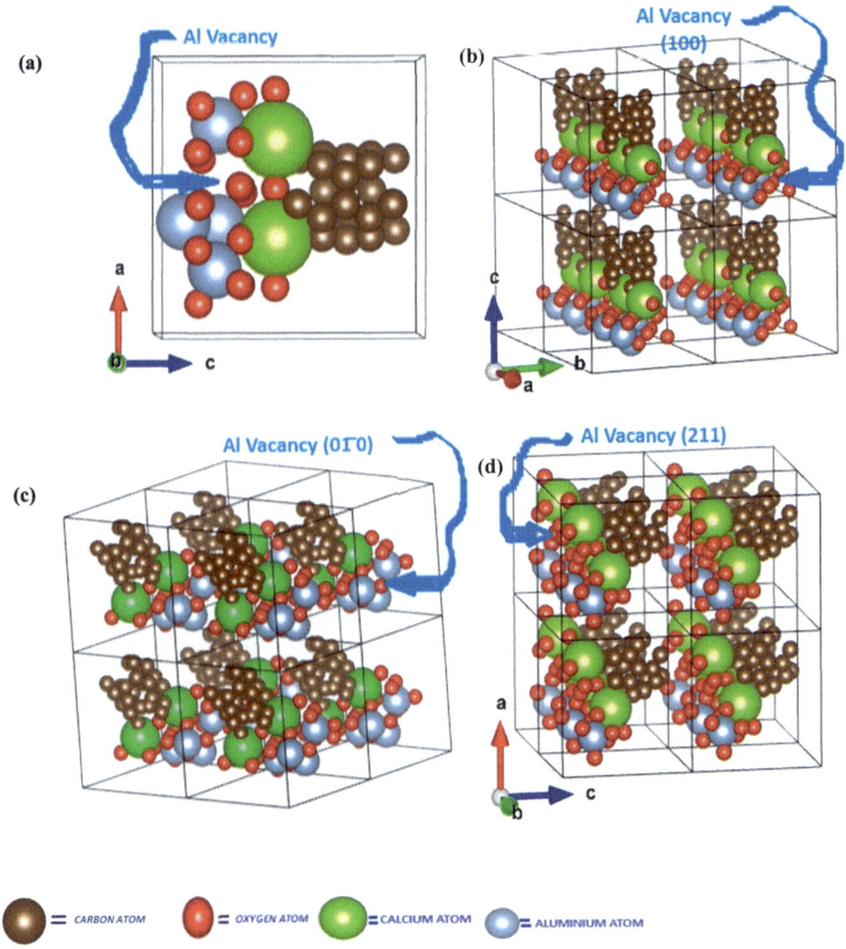

Fig. 2.5 **a** NM Al_2O_3–CaO-CNT(3,3) with aluminium vacancy created via VESTA software with orientation. **b** 3D (100). **c** 3D (0 1 0). **d** 3D (211) [26] (Reused with permission from Ref. [26] Copyright 2024 Elsevier)

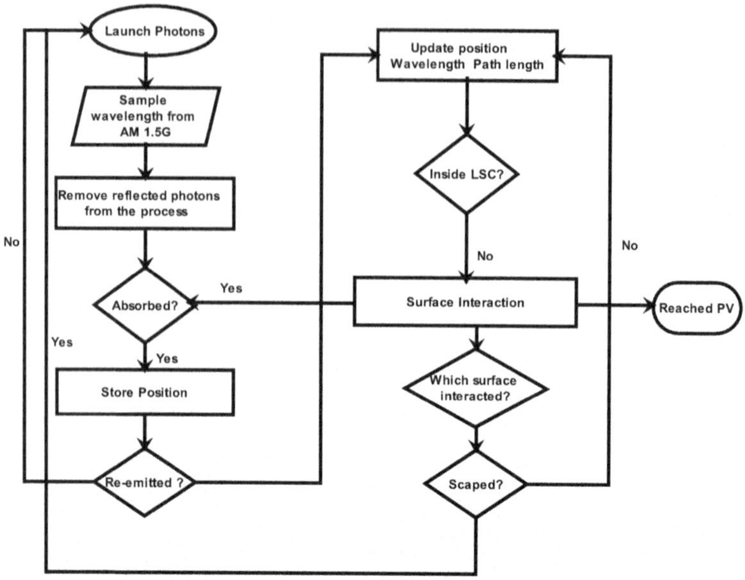

Fig. 2.6 Flowchart of Monte Carlo ray-tracing simulation [27]

energy can be readily transformed into electricity, which is why solar cells are so popular. Luminescent solar concentrators are one way to produce energy at a minimal cost. Their direct integration of solar cells into skyscrapers and crowded metropolitan environments is an advantage. M. R. Mirzaei et al. have used density function theory to model graphene quantum dots. Monte Carlo ray-tracing was used simulation to simulate the gadget (Fig. 2.6). Using Monte Carlo simulation, M. R. Mirzaei et al. have optimized the geometry of the luminous solar concentrator. The final device's energy flux gain has been enhanced by 99%, according to the optimization results. In addition, the destiny of every photon have been determined and examined.

NMs for the delivery of a lenalidomide anticancer drug were studied by Berisha et al. [28] Patients with cancer have been treated mainly with lenalidomide. Up to this moment, graphene and graphene oxide, two common nanostructure models, were used in this study to assess the drug distribution of lenalidomide. To obtain the necessary results of solitary and interacting models, MonC, DenFT, density functional tight binding theory, and molecular dynamic computations were utilized. Stabilized structures and related electronic properties were evaluated for the model systems under study. Formation possibilities were found for both lenalidomide@graphene and lenalidomide@graphene oxide complex models, and aspects of the quantum theory of atoms in the molecule were used to study the associated complex interactions. Changes in frontier molecular orbital levels further indicated that lenalidomide interacted with graphene and graphene oxide structures. The derived conclusions, then, demonstrated that the examined complicated models of lenalidomide@graphene and lenalidomide@graphene oxide were appropriate for executing the intended drug administration procedures utilizing acceptable nano-based surfaces as carriers. In 2022, Dong et al. [29] highlighted the potential of graphite-based NMs as effective conductive additives in cementitious composites with electrical conductivity to advance non-destructive structural health monitoring techniques. Achieving both robust mechanical strength and optimal electrical resistivity in graphite-based NM-reinforced cementitious composites poses a complex multi-objective optimization challenge. This study addresses this multi-objective design optimization problem for graphite-based NM-reinforced cementitious composites by introducing a thorough data-driven strategy that makes use of MAL techniques and the non-dominated sorting genetic algorithm. Initially, using prepared experimental datasets and Bayesian-tuned XGBoost, prediction models are developed for the uniaxial compressive strength and electrical resistivity of these composites. The results demonstrated excellent predictive performance for both properties, with high R-squared values (0.95 and 0.92 for uniaxial compressive strength, 0.99 and 0.98 for electrical resistivity) and low absolute mean error scores (1.24 and 3.44 for uniaxial compressive strength, 0.15 and 0.22 for electrical resistivity). MAL theories further elucidate the influence of key features on graphite-based NM-reinforced cementitious composite properties, aiding in the identification of optimization variables and their constraints for multi-objective design optimization. Subsequently, an NSGA-II-based optimization programme is implemented, using the developed prediction models as objective functions to concurrently optimize the graphite-based NM-reinforced cementitious composites' uniaxial compressive strength and electrical resistivity characteristics. This approach successfully generates a set of Pareto-optimal solutions, facilitating informed parameter selections for designing these composites with desired properties. The intricate microstructures of NMs pose a considerable barrier in forecasting their behaviour under a variety of situations, as noted by Wang et al. [30] in 2024. High-fidelity modelling techniques, such as MolDS, are effective but resource-intensive. A promising method for quickly exploring design options is MAL models. In this work, a deep learning framework was developed based on a conditional generative adversarial network to predict the evolution of grain boundary networks in nano-crystalline materials under mechanical loads, combining both morphological and atomic features. To get precise data, the researchers simulated MolDS on nano-crystalline tungsten. This data was then utilized to train the conditional generative adversarial network model. The model's performance was assessed by contrasting it with a convolutional autoencoder model and examining the ways in which variations in geometric morphology and loading circumstances influenced the model's predictions. In forecasting the evolution of grain boundary networks under different settings, the conditional generative adversarial network model demonstrated a strong level of accuracy. With its potential to forecast the behaviour of various NMs, this framework could provide insightful information for applications in materials science. The results of research on geopolymers (GPs), which are often seen as a low-carbon alternative to building materials but whose intrinsic brittleness limits their use in more general engineering applications, were given in 2023 by Li et al. [31]. The combined impacts of polyethylene fibre (PEF) and nano-calcium carbonate (NCC) on the toughening mechanisms of GP composites were examined in this study by multi-scale AbIn modelling and testing. Due to strong bonding between N/CASH and polyethylene and nanoscale calcium-induced electrostatic attractions, it was discovered that NCC significantly increased the aggregation of GP gels at the micro-scale on the surface of PEF. At the meso-scale, this resulted in enhanced fibre dispersion. In terms of macro-scale bending toughness and multiple cracking behaviour, the combination of 12 and 6 mm PE fibres with 1% nano-calcium carbonate produced the best results. It was discovered that there was a correlation between the fibre reinforcing index and toughness, toughness and crack amount, and toughness and crack fractal size, respectively, with linear, natural logarithmic, and exponential functions. PEF-reinforced GP composites' bending toughness can be seen visually by examining the fractal dimension and quantity of the cracks. Notable results on interfaces in NMs were reported by Muller et al. [32] in 2024. NMs are known to function as ultra-fast diffusion channels because of local structural, chemical, and electronic changes. In particular, relative to its bulk diffusivity, the interface diffusivity of copper in Cu/AlN nano-multilayers can be up to 100 times higher, which may cause interfacial pre-melting of copper. Through extensive AbIn calculations, the researchers investigated the vacancy formation and migration energies in these nano-multilayers to understand the mechanisms behind this rapid interface diffusion. They discovered that both the Al-terminated and N-terminated interfaces enhance atomic mobility at the interface by reducing the vacancy formation and migration energies in the copper layers. Additionally, the energies associated with out-of-plane vacancy migration suggest a strong tendency for vacancies to segregate towards both types of interfaces.

In 2024, Purton et al. [33] addressed the persistent issue of disposing of high-level radioactive waste, highlighting the necessity for methods to assess both the extent and type of damage, and to correlate crystal damage levels with the leaching of radionuclides and the dissolution of the matrix. Several MolDS have been performed to relate the viability of materials for storage matrices to the damage resulting from recoil events. In this work, new insights into the healing process of $Gd_2Ti_2O_7$ following a simulated cascade and the complex diffusion of anions and cations were obtained over long times, up to several seconds, using adaptive kinetic MonC (Fig. 2.7). The reported results, when compared with previous data, indicate that MolDS, which typically cover up to nanoseconds, fail to account for rare high-barrier ionic diffusion events, thus limiting the understanding of radiation damage in materials from short-term MolDS alone.

2.4 Future Prospects

The field of CD in NMs stands at the intersection of theoretical science and practical application, heralding a new era of innovation. At its core, this discipline leverages advanced computational methods to predict and optimize the properties of NMs before their physical synthesis. The essential ideas combine multi-scale

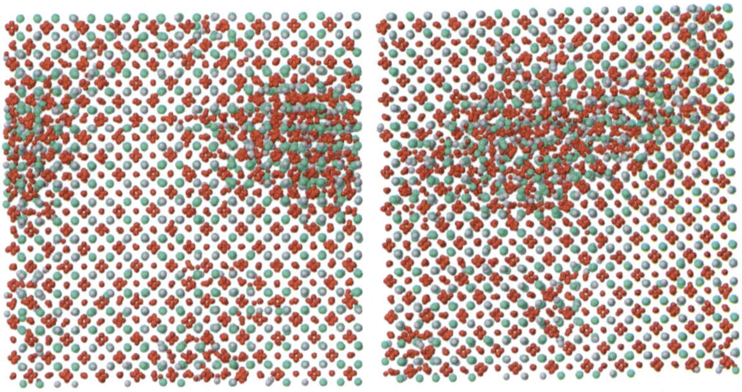

Fig. 2.7 Simulation cells radiation damage, as examined using adaptive kinetic MonC simulations, depicted from two perspectives: along (100) and (010) planes [33] (Reused with permission from Ref. [33] Copyright 2024 Elsevier)

modelling, quantum physics, and MolDS to enable scientists to accurately simulate the behaviour of materials at the atomic and molecular levels. This predictive power not only expedites the process of discovery but also lowers the expenses and duration of experimental trials. The potential for combining artificial intelligence and MAL with CD is enormous. By analysing large datasets, these technologies can find hidden patterns and correlations that improve material characteristics and functionalities even more. Opportunities include the creation of innovative biomaterials for medical uses, next-generation semiconductors, and extremely effective energy storage devices. CD and nanotechnology are coming together to change a number of industries and enable breakthroughs that were previously only possible in science fiction. The future of NMs design is poised to unleash previously unheard-of possibilities through persistent improvements in algorithms, processing capacity, and interdisciplinary cooperation, propelling developments in technology and materials research [34–40].

2.5 Conclusion

To sum up, the application of CD to the study and synthesis of NMs represents an important development in the domains of materials science and engineering. Using a variety of computational methods, such as CGM, MonC simulations, DenFT, MolDS, and MAL, researchers may discover a great deal about the properties and behaviours of NMs. These methods not only serve as a helpful addition to experimental methods but also enhance the efficiency and accuracy of material production and characterization. Future prospects are promising for the field of CD in NMs. Innovation and interdisciplinary collaboration are critical to addressing the complex issues in this field. As computational capabilities grow, so too will the capacity to design and refine novel NMs for a myriad of uses, driving scientific advancements and driving technological innovation. The importance of CD in NMs research is emphasized in this review, which can be very helpful to academics and practitioners in improving their understanding and use of these innovative materials.

References

1. Peng B, Qin J, Li Y, Wu K, Kuang Y, Jiang F (2024) Recent advances in nanomaterials-enabled active food packaging: nanomaterials synthesis, applications and future prospects. Food Control 163:110542. https://doi.org/10.1016/j.foodcont.2024.110542
2. Nyabadza A, Makhesana M, Plouze A, Kumar A, Ramirez I, Krishnamurthy S, Vazquez M, Brabazon D (2024) Advanced nanomaterials and dendrimers in water treatment and the recycling of nanomaterials: a review. J Environ Chem Eng 12:112643. https://doi.org/10.1016/j.jece.2024.112643
3. Liu X, Ba R, Wang W, Zhang Y, Bao B, Chen P, Yao W, Zhu J-J, Zhang L, Cheng F-F (2024) Roles of nanomaterials in thrombin detection. TrAC Trends Analytical Chem 175:117734. https://doi.org/10.1016/j.trac.2024.117734
4. Urmi R, Banerjee P, Singh M, Singh R, Chhillar S, Sharma N, Chandra A, Singh N, Qamar I (2024) Revolutionizing biomedicine: aptamer-based nanomaterials and nanodevices for therapeutic applications. Biotechnol Rep: e00843. https://doi.org/10.1016/j.btre.2024.e00843
5. Wu L, Yuan R, Wen T, Qin Y, Wang Y, Luo X, Liu J-W (2024) Recent advances in functional nucleic acid decorated nanomaterials for cancer imaging and therapy. Biomed Pharmacother 174:116546. https://doi.org/10.1016/j.biopha.2024.116546
6. Yang EC, Divine R, Miranda MC, Borst AJ, Sheffler W, Zhang JZ, Decarreau J, Saragovi A, Abedi M, Goldbach N, Ahlrichs M, Dobbins C, Hand A, Cheng S, Lamb M, Levine PM, Chan S, Skotheim R, Fallas J, Ueda G, Lubner J, Somiya M, Khmelinskaia A, King NP, Baker D (2023) Computational design of non-porous, pH-responsive antibody nanoparticles. https://doi.org/10.1101/2023.04.17.537263
7. He L, Cui Z, Sun X, Zhao J, Wen D (2022) Sensitivity analysis of the catalysis recombination mechanism on nanoscale silica surfaces. Nanomaterials 12:2370. https://doi.org/10.3390/nano12142370
8. Hassoul S, Menouar S, Benseridi H, Cho JR (2021) Dynamical invariant applied on general time-dependent three coupled nano-optomechanical oscillators. J Nanomater 2021:6903563. https://doi.org/10.1155/2021/6903563
9. King NP, Sheffler W, Sawaya MR, Vollmar BS, Sumida JP, Andre I, Gonen T, Yeates TO, Baker D (2012) Computational design of self-assembling protein nanomaterials with atomic level accuracy. Science 336:1171–1174. https://doi.org/10.1126/science.1219364
10. Zhang X, Lu G (2014) Computational design of core/shell nanoparticles for oxygen reduction reactions. J Phys Chem Lett 5:292–297. https://doi.org/10.1021/jz4024699
11. Burello E (2015) Computational design of safer nanomaterials. Environ Sci Nano 2:454–462. https://doi.org/10.1039/C5EN00066A
12. Yang EC, Divine R, Miranda MC, Borst AJ, Sheffler W, Zhang JZ, Decarreau J, Saragovi A, Abedi M, Goldbach N, Ahlrichs M, Dobbins C, Hand A, Cheng S, Lamb M, Levine PM, Chan S, Skotheim R, Fallas J, Ueda G, Lubner J, Somiya M, Khmelinskaia A, King NP, Baker D (2024) Computational design of non-porous pH-responsive antibody nanoparticles. Nat Struct Mol Biol. https://doi.org/10.1038/s41594-024-01288-5
13. Ashwini T, Narayan R, Shenoy PA, Nayak UY (2022) Computational modeling for the design and development of nano based drug delivery systems. J Mole Liquids 368:120596. https://doi.org/10.1016/j.molliq.2022.120596
14. Duncan GA, Bevan MA (2015) Computational design of nanoparticle drug delivery systems for selective targeting. Nanoscale 7:15332–15340. https://doi.org/10.1039/C5NR03691G

15. Shukla S, Jakowski J, Kadian S, Narayan RJ (2023) Computational approaches to delivery of anticancer drugs with multidimensional nanomaterials. Comput Structural Biotechnol J 21:4149–4158. https://doi.org/10.1016/j.csbj.2023.08.010
16. Agarwal S, Banerjee AS (2024) Solution of the Schrödinger equation for quasi-one-dimensional materials using helical waves. J Comput Phys 496:112551. https://doi.org/10.1016/j.jcp.2023.112551
17. Badar MS, Shamsi S, Ahmed J, Alam MA (2022) Molecular dynamics simulations: concept, methods, and applications, transdisciplinarity, 5, 131–151.https://doi.org/10.1007/978-3-030-94651-7_7
18. Chu L, Shi J, de Cursi ES, Xu X, Qin Y, Xiang H (2018) Monte Carlo-based finite element method for the study of randomly distributed vacancy defects in graphene sheets. J Nanomater 2018:3037063. https://doi.org/10.1155/2018/3037063
19. Wang M, Wang T, Cai P, Chen X (2019) Nanomaterials discovery and design through machine learning. Small Methods 3:1900025. https://doi.org/10.1002/smtd.201900025
20. Atilhan M, Aparicio S (2019) Molecular dynamics simulations of mixed deep eutectic solvents and their interaction with nanomaterials. J Mol Liq 283:147–154. https://doi.org/10.1016/j.molliq.2019.03.068
21. Abbaspour M, Jorabchi MN, Akbarzadeh H, Ahmadi N (2021) Molecular dynamics simulation of carbon peapod-like nanomaterials in desalination process. Desalination 504:114975. https://doi.org/10.1016/j.desal.2021.114975
22. Borówko M, Staszewski T (2024) Molecular dynamics simulations of different nanoparticles at substrates. Int J Mol Sci 25:4550. https://doi.org/10.3390/ijms25084550
23. Das DK, Kumar B (2024) Molecular dynamics simulation study to evaluate mechanical properties of plumbene using bending, oscillation and equilibrium MD approaches. Comput Mater Sci 233:112678. https://doi.org/10.1016/j.commatsci.2023.112678
24. Adara PP, Oyinbo ST, Jen T-C (2023) Density functional theory simulation and modeling of the electrical and mechanical properties of Al_2O_3–CAO-CNT(3,3) nanomaterial. Comput Mater Sci 218:111939. https://doi.org/10.1016/j.commatsci.2022.111939
25. Bassey VM, Okon EED, Louis H, Benjamin I, Chukwuemeka K, Gber TE, Ezekiel MC, Qader SW, Adeyinka AS (2023) Nano-enhanced drug delivery of dacarbazine using heteroatoms (B, N, P, S) doped Ag-functionalized silicene nanomaterials: insight from density functional theory. Chem Phys Impact 7:100297. https://doi.org/10.1016/j.chphi.2023.100297
26. Adara PP, Oyinbo ST, Jen T-C (2024) Electrical and mechanical properties variation of Al_2O_3–CaO-CNT(3,3) nanomaterial due to Al vacancy and temperature: DFT approach. Res Mater 2:100571
27. Mirzaei MR, Rostami A, Matloub S, Nazari M (2021) Design and optimization of graphene quantum dot-based luminescent solar concentrator using Monte-Carlo simulation. Energy Built Environ 4:140–147. https://doi.org/10.1016/j.enbenv.2021.10.002
28. Berisha A (2023) Density functional theory and quantum mechanics studies of 2D carbon nanostructures (graphene and graphene oxide) for Lenalidomide anticancer drug delivery. Comput Theor Chem 1230:114371. https://doi.org/10.1016/j.comptc.2023.114371
29. Dong W, Huang Y, Lehane B, Ma G (2022) Multi-objective design optimization for graphite-based nanomaterials reinforced cementitious composites: a data-driven method with machine learning and NSGA-II. Constr Build Mater 331:127198. https://doi.org/10.1016/j.conbuildmat.2022.127198
30. Wang Y, Kazemi A, Jing T, Ding Z, Li L, Yang S (2024) Rapid prediction of grain boundary network evolution in nanomaterials utilizing a generative machine learning approach. Extreme Mech Lett 70:102172. https://doi.org/10.1016/j.eml.2024.102172
31. Li L, Ma Z, Ming X (2023) Multiscale ab-initio modeling and experiment of nano-$CaCO_3$ and fiber synergy on toughening low-carbon geopolymer composites. Mater Des 233:112280. https://doi.org/10.1016/j.matdes.2023.112280
32. Muller Y, Antušek A, Jeurgens LPH, Turlo V (2024) Anomalously low vacancy formation energies and migration barriers at Cu/AlN interfaces from ab initio calculations. Scripta Mater 248:116126. https://doi.org/10.2139/ssrn.4731302

33. Purton JA, Hull S, Teobaldi G (2024) Adaptive kinetic Monte Carlo analysis of healing mechanisms in radiation damaged pyrochlore $Gd_2Ti_2O_7$. Acta Mater 267:119726. https://doi.org/10.1016/j.actamat.2024.119726
34. Yang RX, McCandler CA, Andriuc O, Siron M, Woods-Robinson R, Horton MK, Persson KA (2022) Big data in a nano world: a review on computational, data-driven design of nanomaterials structures, properties, and synthesis. ACS Nano 16:19873–19891. https://doi.org/10.1021/acsnano.2c08411
35. Balasubramanian G (2023) Advances in nanomaterials: fundamentals, properties and applications. https://doi.org/10.1007/978-3-319-64717-3
36. Ramazani A (2019) Computational materials design for renewable energy applications. Nanomaterials 9:1–25. https://doi.org/10.3390/nano9010001
37. Giannazzo F, Shtepliuk I (2023) Computational design of gas sensors based on V_3S_4 monolayer. Nanomaterials 13:774. https://doi.org/10.3390/nano13030774
38. Bystrov VS (2023) Special issue: simulation and modeling of nanomaterials. Nanomaterials 14:50–66. https://doi.org/10.3390/nano14010050
39. Hu S, Das SR, Monshat H (2023) 2D materials: applications for electrochemical energy storage devices. In: Advances in nanomaterials: fundamentals, properties and applications, 61–75. https://doi.org/10.1016/B978-0-12-820000-0.00006-4
40. Sanchez O, Kim JM, Balasubramanian G (2023) Graphene analogous elemental van der Waals structures. In: Advances in nanomaterials: fundamentals, properties and applications, 77–93. https://doi.org/10.1016/B978-0-12-820000-0.00007-6

Chapter 3
Computational Modeling of Radiative Magneto Maxwell Nanofluid: An Application to Improving the Efficiency of Energy Conversion and Storage Systems

Pudhari Srilatha, M. C. Jayaprakash, Konduru Sarada, R. J. Punith Gowda, and R. Naveen Kumar

Abstract Researchers in the field of heat transfer continually look for novel approaches to improve the efficiency of energy devices through heat transfer. Various working conditions and liquids have been tested to improve the heat transport method. The interface of such liquid into the prevailing system may be suitable for reducing capital costs, improving the working efficacy, and better designing the anticipated system. The utilization of Maxwell fluid combined with nanoparticle suspension has great potential in improving the efficiency of energy conversion and storage systems. There is a possibility that they could be utilized in complex cooling systems for power-producing facilities, which would increase the total energy efficiency. The current study examines the significance of heat source and sink on the Maxwell nanoliquid stream past an oscillatory stretchy surface. Also, the impact of the magnetic field on liquid flow is considered. The governing partial differential equations (PDEs) are nondimensionalized with similarity variables. The obtained dimensionless equations are also solved using the finite difference method (FDM). Graphical illustrations

P. Srilatha
Department of Mathematics, Institute of Aeronautical Engineering, Hyderabad 500043, India

M. C. Jayaprakash
Department of Information Technology, University of Technology and Applied Sciences, Al Musanna, Sultanate of Oman

K. Sarada
Department of Mathematics, Government City College, Nayapul, Hyderabad 500002, India

R. J. Punith Gowda (✉)
Department of Mathematics, Bapuji Institute of Engineering and Technology, Davanagere, Karnataka 577004, India
e-mail: rjpunithgowda@gmail.com

R. N. Kumar
Department of Mathematics, Amrita School of Engineering, Amrita Vishwa Vidyapeetham, Bengaluru, Karnataka, India

R. Naik et al. (eds.), *Advanced Nanomaterials for Energy Storage Devices*, Nanostructure Science and Technology, https://doi.org/10.1007/978-3-031-74730-4_3

show the impact of numerous factors on the velocity, temperature, and concentration profiles. Results reveal that the upsurge in the heat source/sink parameter increases the thermal profile. Elevated heat source/sink parameter values can potentially modify the fluid's thermal relaxation characteristics, impacting the temperature's progressive evolution. As the values of the ratio of oscillating frequency to stretching rate increases, the thermal profile decreases because the fluid undergoes more intensive mixing when subjected to rapid oscillations, which promotes heat dispersion and decreases the fluid's temperature.

Keywords Maxwell nanofluid · Oscillatory stretchy surface · Magnetic field · Heat source/sink · Chemical reaction

Nomenclature

Symbol

Sc	Schmidt number
t	Dimensionless time
ρ	Density
μ	Dynamic viscosity
(u, v)	Velocity components
$f(y, t)$	Dimensionless velocity
$(\overline{x}, \overline{y})$	Axis
Q_0	Heat source/sinkXE ""
k_r	Chemical reaction
T_∞	Ambient temperature
$\chi(y, t)$	Dimensionless concentration
Q_1	Heat source/sink
Nu	Nusselt number
Γ_1	Maxwell parameter
k_c	Chemical reaction parameter
ω	Angular velocity
B_0	Magnetic field
D_B	Diffusion coefficient
T	Temperature
k	Thermal conductivity
Pr	Prandtl number
ν	Kinematic viscosity
c_p	Specific heat
M	Magnetic parameter
$\theta(y, t)$	Dimensionless temperature
Re	Reynolds number

S	The ratio of oscillating frequency to stretching rate
b	Stretching rate
Sh	Sherwood number
C	Concentration
τ	Time
Cf	Skin friction

Subscripts

nf	Nanofluid
∞	Ambient
f	Fluid
w	Wall

3.1 Introduction

The boundary layer flows of non-Newtonian liquids (NNLs) are receiving much attention due to their practical uses in manufacturing and industrial processes. Examples of such applications include the production of plastic polymers, hot rolling paper, optical fibers, metal spinning, drilling muds, and cooling metallic plates in cooling baths. Because no one model can accurately predict every characteristic of non-Newtonian materials, researchers have previously put forward a variety of non-Newtonian models. The three foremost groups these models fall into are rate-, differential-, and integral-type liquids. The considered liquid model is a subclass of a liquid recognized as a Maxwell fluid, a rate-type liquid. This liquid model projects the impact of relaxation time. Fluids of the differential kind are not able to foresee such consequences. The magnetic discipline-precipitated Maxwell nanofluid movement throughout parallel rotating disks was inspected by Ahmadian et al. [1]. The magnetic field impact on the Maxwell nanoliquid flow via a surface was deliberated by Jamshed [2]. Kumar et al. [3] assessed the Maxwell nanofluid over the various carbon nanotubes under the impact of a magnetic dipole. Gowda et al. [4] deliberated the Casson–Maxwell nanofluid's slip flow restricted to two stretchable sheets. Khan et al. [5] examined the Maxwell nanoliquid flow past a stretchy sheet by contemplating heat radiation and chemical interactions.

Due to its many impressive practical applications in various engineering and manufacturing procedures, boundary layer movement analysis with combined mass and heat transport across a stretchy sheet has attracted the consideration of many investigators and engineers in recent years. However, there are circumstances in which the sheet is stretched and experiences periodic oscillations within its plane. This time-dependent flow phenomenon allows different aspects of liquid motion to be examined at diverse times. Nawaz et al. [6] inspected the Darcy–Forchheimer stream of an NNL past an oscillatory sheet in the magnetohydrodynamic (MHD) effect. The evaluation of a curvilinear hybrid nanoliquid movement past a curved oscillatory stretchy surface with nonlinear thermal radiation was studied by Imran et al. [7]. Arif et al. [8] analyzed the movement of non-Newtonian fluid via an oscillatory sheet. The thermal transport behavior of a micropolar liquid over an oscillating moving sheet in a mixed convective Darcy–Forchheimer flow was probed by Nawaz et al. [9]. Alqurashi et al. [10] deliberated the movement of a micropolar nanofluid past an oscillating stretchy surface with the influence of bioconvection.

There are several uses for examining magnetic field effects in engineering, chemistry, and physics. Recent years have seen the utilization of hydro-magnetic processes in metallurgy and the polymer sector, namely in the stretching of plastic sheets. To be more precise, it should be noted that continuous strips or filaments are often cooled via the drawing process using a quiescent fluid and that stretching of the strips occurs during this process. Copper wire drawing, annealing, and thinning may be mentioned. In each of these situations, the intended final product characteristic and the pace at which these strips cool down when drawn through an electrically leading liquid in the incidence of a magnetic field determine the final product's characteristics to a large degree. The magnetic discipline-precipitated micropolar fluid stream past a stretched sheet was scrutinized by Patel et al. [11]. The movement of fluids across a spinning sphere with the influence of a torsional magnetic subject was evaluated by Sharma and Srivastava [12]. Saini et al. [13] discussed the importance of the magnetic field and entropy evaluation on the liquid flow over the surface. Maurya et al. [14] investigated the impact of a magnetic field on stress liquid around a spherical solid in a permeable media. Gogoi et al. [15] looked at the effects of a homogeneous magnetic field on liquid mobility over a stretchable surface.

Many issues may have discernible temperature variance between the surrounding liquid and the surface. This forces one to consider temperature-dependent heat sources or sinks since they might significantly impact the aspects of heat transport. Although its strength is stimulating to estimate internal heat production or absorption accurately, most physical circumstances can be expressed by simple numerical models that represent the average behavior of the system. Thus, consideration of the external heat production/absorption parameter or heat source/sink (HSS) is necessary. Dawar et al. [16] focused on the effect of velocity slip on fluid flow passing a surface while HSS is present. The flow of liquid past a conducting channel with the significance of HSS and Hall effects was probed by Kumar et al. [17]. Utilizing the heat flux model, Alqawasmi et al. [18] studied the flow of ternary nanoliquid stream past a disk with the HSS. The impact of HSS on heat transport analysis of the convective fluid flow past a stretchy sheet in the presence of a magnetic field was scrutinized by Sharma and Badak [19]. Manjunatha et al. [20] examined the impact of HSS on the fluid flow past a permeable surface.

For its wide range of manufacturing applications, chemical reactions play a significant role in heat and mass transfer research in science and engineering technology. Due to its crucial role in developing chemical processing machinery, the synthesis of polymers, cooling towers, pollution, the formation and dispersion of fog, temperature, and fiber insulation, among other things, it is deemed necessary. The purpose of industrial chemical practices is to convert raw materials into advanced standard products through chemical reactions, such as oxidation and condensation of components, cold-induced crop destruction, food processing, and liquid dispersal on fields of crops. Reactors are used for these kinds of chemical reactions. The concentration and reaction rate are directly correlated in a first-order chemical reaction. Several writers have thought about and developed their findings on chemical reaction impacts on the process of mass and heat with various geometries in light of the relevant applications. The impact of chemical reactions on the liquid stream past a surface was examined by Kumar et al. [21]. The Casson nanofluid flow over a stretched, curved surface in the presence of a magnetic field and a chemical reaction was studied by Kumar et al. [22]. Eric et al. [23] studied the dual-phase flow of turbulent fluid inside a pipe with chemical reactions. Das and Begum [24] analyzed the consequences of the binary chemical process, nonlinear convective thermal condition, and Arrhenius activation energy in Casson fluid flow. Farooq et al. [25] deliberated the chemically reactive fluid movement over an inclined stretchable sheet.

It is evident from the literature survey that no study has been described about the Maxwell nanoliquid stream across an oscillating stretchable sheet in the incidence of a magnetic field. To address this research gap, the current work examines the effects of HSS and chemical reactions on the Maxwell nanofluid flow via an oscillating stretchable surface while a magnetic field is present. Scientists and engineers may improve the efficacy of chemical reactions or heat transfer by scheming systems with optimum flow and researching how reactions affect flow. Similarity variables are used to convert the PDEs into their dimensionless form. The FDM is also used to solve the resultant equations numerically. A graphical depiction is provided to demonstrate the impact of several parameters on the profiles of concentration, temperature, and velocity.

3.2 Mathematical Formulation

Let us consider the movement of a Maxwell nanofluid in two dimensions over an oscillatory stretchy sheet at $\overline{y} = 0$ in the existence of a magnetic field. Also, the impact of chemical reaction and HSS on the nanofluid flow is considered. The stretchable surface is followed by the $\overline{x}$-axis in this case, and the $\overline{y}$-axis is taken normal to it as depicted in Fig. 3.1. This sheet is supposed to oscillate harmonically in its plane with velocity $u_w = b\overline{x} \sin \omega \tau$ in which b is the stretching rate, τ is the time, and ω signifies the angular velocity. The liquid filled the space at $\overline{y} > 0$. Let (u, v) be the velocity components in the $(\overline{x}, \overline{y})$ directions and (T, C) are the temperature and concentration. The governing equations of the problem are mentioned as follows (see [22, 26]).

$$\frac{\partial v}{\partial \overline{y}} + \frac{\partial u}{\partial \overline{x}} = 0, \tag{3.1}$$

$$u\frac{\partial u}{\partial \overline{x}} + \frac{\partial u}{\partial \tau} + v\frac{\partial u}{\partial \overline{y}} = \nu_{nf}\frac{\partial^2 u}{\partial \overline{y}^2} - \Gamma\left(v^2\frac{\partial^2 u}{\partial \overline{y}^2} + u^2\frac{\partial^2 u}{\partial \overline{x}^2} + 2uv\frac{\partial^2 u}{\partial \overline{x}\partial \overline{y}}\right) - \frac{\sigma_{nf}}{\rho_{nf}}B_0^2 u, \tag{3.2}$$

$$\frac{\partial T}{\partial \tau} + u\frac{\partial T}{\partial \overline{x}} + v\frac{\partial T}{\partial \overline{y}} = \frac{k_{nf}}{(\rho c_p)_{nf}}\frac{\partial^2 T}{\partial \overline{y}^2} + \frac{Q_0}{(\rho c_p)_{nf}}(T - T_\infty), \tag{3.3}$$

$$\frac{\partial C}{\partial \tau} + u\frac{\partial C}{\partial \overline{x}} + v\frac{\partial C}{\partial \overline{y}} = -k_r(C - C_\infty) + D_B\frac{\partial^2 C}{\partial \overline{y}^2} \tag{3.4}$$

The corresponding boundary conditions are specified as follows,

$$\begin{gathered} u = u_w = b\overline{x}\sin\omega\tau, v = 0, C = C_w, T = T_w \quad \text{at} \quad \overline{y} = 0, \tau > 0, \\ u = 0, T \to T_\infty, C \to C_\infty \quad \text{as} \quad \overline{y} \to \infty \end{gathered} \tag{3.5}$$

The similarity variables are presented as follows,

$$\begin{gathered} y = \sqrt{\frac{b}{\nu_f}}\overline{y}, t = \omega\tau, u = b\overline{x}f_y(y,t), v = -\sqrt{b\,\nu_f}f(y,t), \\ \chi(y,t) = \frac{C - C_\infty}{C_w - C_\infty}, \theta(y,t) = \frac{T - T_\infty}{T_w - T_\infty} \end{gathered} \tag{3.6}$$

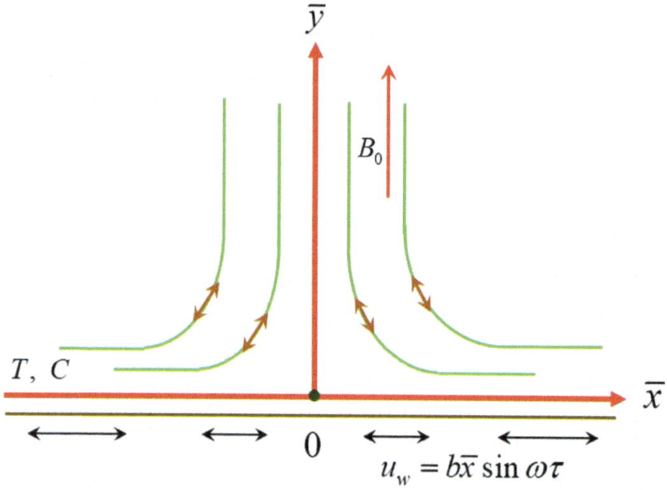

Fig. 3.1 Flow geometry

With the aid of Eqs. (3.6), (3.1) is satisfied and Eqs. (3.2–3.4) are reduced as follows,

$$\frac{\nu_{nf}}{\nu_f} f_{yyy} + f f_{yy} - (f_y)^2 - S f_{yt} - \Gamma_1 \left(f^2 f_{yyy} - 2f f_y f_{yy}\right) - \frac{\sigma_{nf}}{\sigma_f} \frac{\rho_f}{\rho_{nf}} M f_y = 0, \tag{3.7}$$

$$\frac{1}{\Pr} \frac{k_{nf}}{k_f} \frac{(\rho c_p)_f}{(\rho c_p)_{nf}} \theta_{yy} + f\theta_y - S\theta_t + \frac{(\rho c_p)_f}{(\rho c_p)_{nf}} Q_1 \theta = 0, \tag{3.8}$$

$$\frac{1}{Sc} \chi_{yy} + f \chi_y - S \chi_t - k_c \chi = 0, \tag{3.9}$$

Equation (3.5) is reduced as,

$$\begin{gathered} f_y(0,t) = \sin t, f(0,t) = 0, \theta(0,t) = 1, \chi(0,t) = 1, \\ f_y(\infty,t) = 0, \theta(\infty,t) = 0, \chi(\infty,t) = 0 \end{gathered} \tag{3.10}$$

where,

$\Gamma_1 = \Gamma b$	(Maxwell parameter),
$\Pr = \frac{\nu_f (\rho c_p)_f}{k_f}$	(Prandtl number),
$\mathrm{Sc} = \frac{\nu_f}{D_B}$	(Schmidt number),
$S = \frac{\omega}{b}$	(Unsteadiness parameter),
$M = \frac{\sigma_f}{\rho_f} \frac{B_0^2}{b}$	(Magnetic parameter),
$k_c = \frac{k_r}{b}$	(Chemical reaction parameter),
$Q_1 = \frac{Q_0}{(\rho c_p)_f b}$	(Heat source/sink parameter), and
$t = \omega\tau$	(Dimensionless time)

The physical quantities are obtained as,

$$C_{fx} = \frac{1}{\mathrm{Re}^{1/2}} f_{yy}(0,t), \frac{\mathrm{Nu}_x}{\mathrm{Re}^{1/2}} = -\theta_y(0,t), \frac{\mathrm{Sh}_x}{\mathrm{Re}^{1/2}} = -\chi_y(0,t) \tag{3.11}$$

Thermophysical properties of density, dynamic viscosity, thermal conductivity, and electrical conductivity are expressed as follows (see [28]),

$$\mu_{\mathrm{nf}} = \frac{\mu_f}{(1-\phi)^{2.5}}, \tag{3.12}$$

$$\rho_{\mathrm{nf}} = (1-\phi)\rho_f + \phi\rho_s, \tag{3.13}$$

$$(\rho c_p)_{\mathrm{nf}} = (1-\phi)(\rho c_p)_f + \phi(\rho c_p)_s, \tag{3.14}$$

$$\frac{k_{\mathrm{nf}}}{k_f}=\frac{k_s-2\phi\left(k_f-k_s\right)+2k_f}{k_s+\phi\left(k_f-k_s\right)+2k_f}, \tag{3.15}$$

$$\frac{\sigma_{\mathrm{nf}}}{\sigma_f}=1+\frac{3\left(\frac{\sigma_s}{\sigma_f}-1\right)\phi}{\left(\frac{\sigma_s}{\sigma_f}+2\right)-\left(\frac{\sigma_s}{\sigma_f}-1\right)\phi}. \tag{3.16}$$

3.3 Numerical Procedure

One of the early approaches used for solving differential equations numerically was the finite difference method (FDM), which is specifically utilized for solving the differential form of the governing equations. The method employs a Taylor series expansion to discrete the derivatives of the sequence of variables. The issue's relevant expressions (PDEs) are dimensionless by choosing appropriate variables. Utilizing FDM, these PDEs are numerically solved. The dimensionless PDEs (3.7)–(3.9) are converted to forward difference equations as follows,

Equation (3.7) becomes,

$$\begin{aligned}&\frac{\nu_{\mathrm{nf}}}{\nu_f}\frac{A^n_{i,j+3}-3A^n_{i,j+2}+3A^n_{i,j+1}-A^n_{i,j}}{H_j^3}+A^n_{i,j}\left(\frac{A^n_{i,j+2}+A^n_{i,j}-2A^n_{i,j+1}}{H_j^2}\right)\\&-\left(\frac{A^n_{i,j+1}-A^n_{i,j}}{H_j}\right)^2-S\frac{A^{n+1}_{i+1,j}-A^n_{i,j+1}-A^{n+1}_{i,j}+A^n_{i,j}}{H_iH_n}-\frac{\sigma_{\mathrm{nf}}}{\sigma_f}\frac{\rho_f}{\rho_{\mathrm{nf}}}M\frac{A^n_{i,j+1}-A^n_{i,j}}{H_j}\\&-\Gamma_1\left(\left(A^n_{i,j}\right)^2\frac{A^n_{i,j+3}-3A^n_{i,j+2}+3A^n_{i,j+1}-A^n_{i,j}}{H_j^3}\right.\\&\left.\qquad-2A^n_{i,j}\frac{A^n_{i,j+1}-A^n_{i,j}}{H_j}\left(\frac{A^n_{i,j+2}+A^n_{i,j}-2A^n_{i,j+1}}{H_j^2}\right)\right)=0,\end{aligned} \tag{3.17}$$

Equation (3.8) becomes,

$$\begin{aligned}&\frac{1}{\Pr}\frac{k_{\mathrm{nf}}}{k_f}\frac{\left(\rho c_p\right)_f}{\left(\rho c_p\right)_{\mathrm{nf}}}\left(\frac{Z^n_{i,j+2}+Z^n_{i,j}-2Z^n_{i,j+1}}{H_j^2}\right)+A^n_{i,j}\left(\frac{-Z^n_{i,j}+Z^n_{i,j+1}}{H_j}\right)\\&-S\frac{Z^{n+1}_{i,j}-Z^n_{i,j}}{H_n}+\frac{\left(\rho c_p\right)_f}{\left(\rho c_p\right)_{\mathrm{nf}}}Q_1Z^n_{i,j}=0,\end{aligned} \tag{3.18}$$

Equation (3.9) becomes,

$$\frac{1}{\mathrm{Sc}}\frac{\Phi^n_{i,j+2}-2\Phi^n_{i,j+1}+\Phi^n_{i,j}}{H_j^2}+A^n_{i,j}\frac{\Phi^n_{i,j+1}-\Phi^n_{i,j}}{H_j}$$
$$-S\frac{\Phi^{n+1}_{i,j}-\Phi^n_{i,j}}{H_n}-k_c\,\Phi^n_{i,j}=0, \tag{3.19}$$

Also, Eq. (3.10) is obtained as,

$$\frac{A^n_{i,1}-A^n_{i,0}}{H_0}=\sin t,\, A^n_{i,0}=0,\, Z^n_{i,0}=1,\, \Phi^n_{i,0}=1,$$
$$\frac{A^n_{i,\infty}-A^n_{i,\infty}}{H_\infty}=0,\, Z^n_{i,\infty}=0,\, \Phi^n_{i,\infty}=0. \tag{3.20}$$

3.4 Results and Discussion

A mathematical description of the influence of chemical reaction on magnetically triggered Maxwell nanoliquid flow across an oscillatory stretching sheet is given. Numerical calculations comprehend how various factors affect the thermal, concentration, and velocity fields. We found the required outcomes for the viscous fluid in the graphs and highlighted the numerous parameters affecting the momentum, concentration, and temperature profiles. Thermophysical values of the nanoparticle and base fluid are displayed in Table 3.1.

Figure 3.2 displays the influence of Γ_1 on the time-series of $f_y(y,t)$ with respect to time. Here, the amplitude of the velocity decays with the rise in Γ_1. The reason for this is because the relaxation time found in Γ_1 gives the fluid's viscoelastic qualities, which in turn offer flow resistance. Figure 3.3 demonstrates the impact of M on the velocity profile $f_y(y,t)$. Here, increasing M results in the decline of the velocity profile. Changes in the M might have an impact on how the liquid moves. The fluid particles' velocity and distribution within the flow field may be influenced by the forces applied by the magnetic field. Because Maxwell fluids are viscoelastic,

Table 3.1 Thermophysical values of nanoparticle and base fluid (see [27])

Properties	$\rho\,(\mathrm{kgm}^{-3})$	$c_p(\mathrm{Jkg}^{-1}\,\mathrm{K}^{-1})$	$k(\mathrm{Wm}^{-1}\,\mathrm{K}^{-1})$	$\sigma\,(\mathrm{Sm}^{-1})$
Methanol	792	2545	0.2035	0.5×10^{-6}
Cobalt ferrite ($CoFe_2O_4$)	4907	700	3.7	5.51×10^{9}

they behave in a time-dependent manner. Variations in the velocity profile over time may be caused by the increasing impact of viscoelastic processes as the M rises. Figure 3.4 displays the impact of S on the time-series of velocity profile $f_y(y, t)$. Here, cumulative S results in deterioration of $f_y(y, t)$. An upsurge in S is shown to cause a noticeable phase shift and upsurge in oscillation amplitude which results in a reduction of $f_y(y, t)$.

The influence of Q_1 on $\theta(y, t)$ is displayed in Fig. 3.5. Here, the rise in Q_1 upsurges the $\theta(y, t)$. More heat is supplied to or removed from the liquid per unit volume or per unit time when the heat source/sink parameter is bigger. This directly impacts the fluid's internal thermal energy distribution. Variations in the heat source/sink parameter cause the liquid's temperature gradient to alter. The thermal gradient may steepen or flatten in response to adding or removing more heat. Because of their viscoelastic behavior, Maxwell fluids have distinct relaxation durations for mechanical and thermal processes. Elevated Q_1 have the potential to modify the fluid's

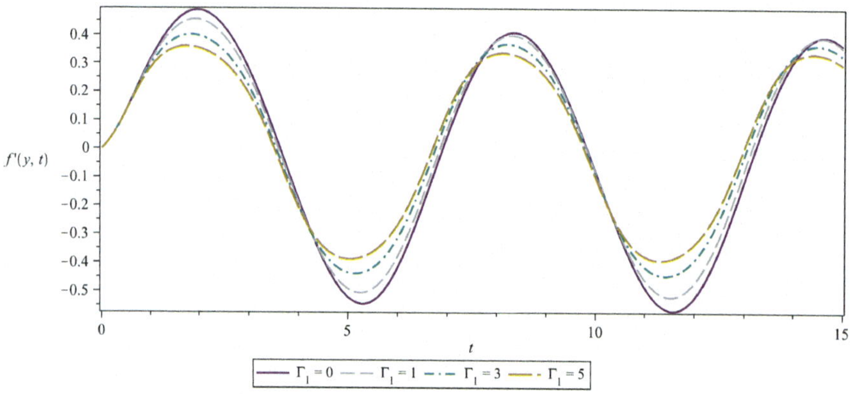

Fig. 3.2 Impact of Γ_1 on $f_y(y, t)$

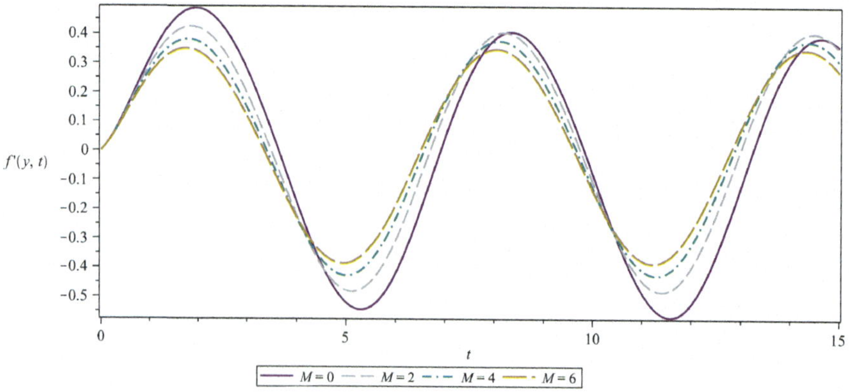

Fig. 3.3 Impact of M on $f_y(y, t)$

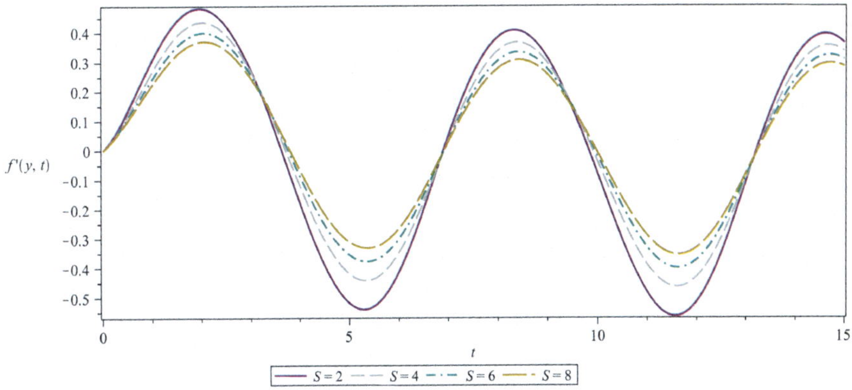

Fig. 3.4 Impact of S on $f_y(y, t)$

thermal relaxation characteristics, hence impacting the temperature's progressive evolution. Figure 3.6 displays the impact of S on $\theta(y, t)$. Here, cumulative S results in the decline of $\theta(y, t)$. A larger ratio of oscillation frequency to stretching rate indicates that sheet oscillation frequency is higher than stretching rate. This might lead to better fluid mixing in the area around the sheet. When subjected to rapid oscillations, the fluid undergoes more intensive mixing, which promotes heat dispersion and decreases the fluid's temperature, which tends to decline in $\theta(y, t)$.

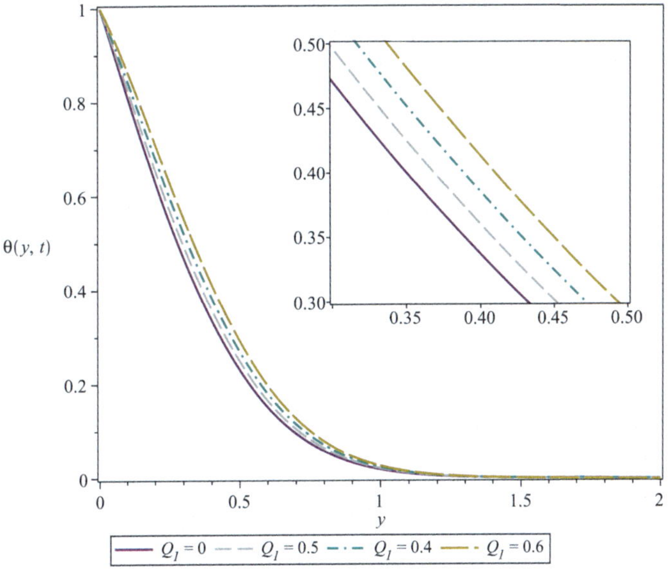

Fig. 3.5 Impact of Q_1 on $\theta(y, t)$

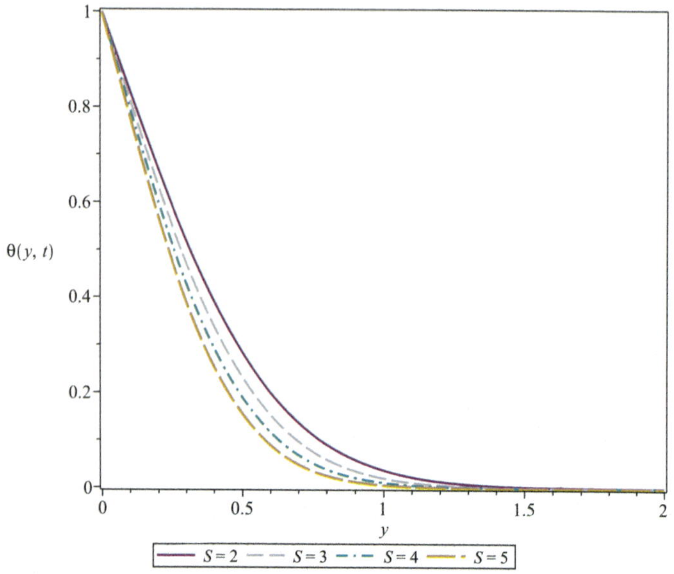

Fig. 3.6 Impact of S on $\theta(y, t)$

Figure 3.7 explores the influence of K_c on $\chi(y, t)$. Here, the development in values of K_c reduces the $\chi(y, t)$. Increasing K_c modifies the kinetics of the chemical processes. The equilibrium concentrations of the reactant and product are often altered as a result. A higher reaction rate parameter might result in a more noticeable reactant depletion, lowering the $\chi(y, t)$. Figure 3.8 displays the impact of S on $\chi(y, t)$. Here, growing S results in the decline of the $\chi(y, t)$. The sheet oscillation frequency may be higher than the stretching rate if the ratio of oscillation frequency to stretching rate is larger. This near the sheet can lead to more mixing. The fluid mixes more vigorously when there are rapid oscillations, which promotes solute dispersion and reduces concentration gradients. Hence, the concentration profile so often tends to decrease.

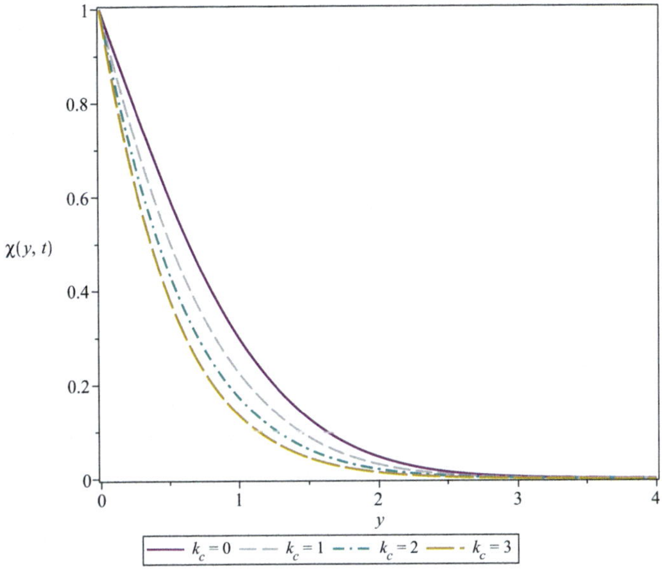

Fig. 3.7 Impact of k_c on $\chi(y, t)$

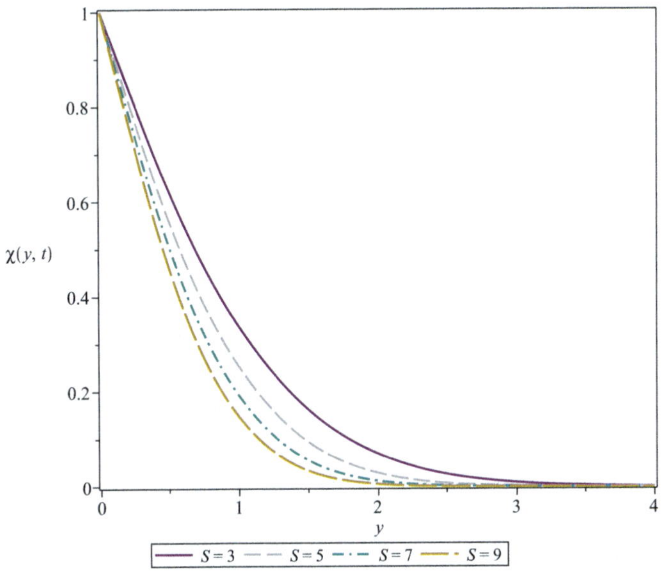

Fig. 3.8 Impact of S on $\chi(y, t)$

3.5 Conclusion

The present analysis inspects the significance of chemical reaction and HSS on the Maxwell nanoliquid stream past an oscillatory stretchy sheet in the presence of the magnetic field. With similarity variables, the governing PDEs are converted to dimensionless form. Furthermore, the numerical approach FDM is used to resolve the derived equations. Graphs illustrate the effects of different factors on the temperature, velocity, and concentration profiles. The main findings of the current investigation are outlined below:

- The influence of Γ_1, S, and M on the velocity profile with respect to time is represented. Here, the velocity amplitude decays with the rise in values of Γ_1, S, and M.
- The upsurge in the HSS parameter raises the thermal profile. Elevated values of the HSS parameter can potentially modify the fluid's thermal relaxation characteristics, impacting the temperature's progressive evolution.
- As the values of S increase, the thermal profiles decline because the fluid undergoes more intensive mixing when subjected to rapid oscillations, which promotes heat dispersion and decreases the fluid's temperature.
- The concentration profile is lowered as the chemical reaction parameter increases. A higher reaction rate parameter might result in a more noticeable reactant depletion, lowering the $\chi(y, t)$.
- As the value of S intensifies, concentration profiles diminish. It is due to fluid mixing more vigorously when there are rapid oscillations, which promotes solute dispersion and reduces concentration gradients.

References

1. Ahmadian A, Bilal M, Khan MA, Asjad MI (2020) The non-Newtonian maxwell nanofluid flow between two parallel rotating disks under the effects of magnetic field. Sci Rep 10(1):17088. https://doi.org/10.1038/s41598-020-74096-8
2. Jamshed W (2021) Numerical investigation of MHD impact on Maxwell nanofluid. Int Commun Heat Mass Transf 120:104973. https://doi.org/10.1016/j.icheatmasstransfer.2020.104973
3. Kumar V, Madhukesh JK, Jyothi AM, Prasannakumara BC, Ijaz Khan M, Chu Y-M (2021) Analysis of single and multi-wall carbon nanotubes (SWCNT/MWCNT) in the flow of Maxwell nanofluid with the impact of magnetic dipole. Comput Theor Chem 1200:113223. https://doi.org/10.1016/j.comptc.2021.113223
4. Gowda RJP, Rauf A, Naveen Kumar R, Prasannakumara BC, Shehzad SA (2022) Slip flow of Casson–Maxwell nanofluid confined through stretchable disks. Indian J Phys 96(7):2041–2049. https://doi.org/10.1007/s12648-021-02153-7
5. Khan KA et al (2024) Exploring the numerical simulation of Maxwell nanofluid flow over a stretching sheet with the influence of chemical reactions and thermal radiation. Results Phys 60:107635. https://doi.org/10.1016/j.rinp.2024.107635

6. Nawaz Y, Arif MS, Abodayeh K (2023) Unconditionally stable numerical scheme for heat transfer of mixed convective Darcy–Forchheimer flow of micropolar fluid over oscillatory moving sheet. J Comput Non-lin Dyn 18(041006). https://doi.org/10.1115/1.4056969
7. Imran M, Naveed M, Iftikhar B, Abbas Z (2023) Heat transfer analysis in a curvilinear flow of hybrid nanoliquid across a curved oscillatory stretched surface with non-linear thermal radiation. ZAMM J Appl Math Mech Z Für Angew Math Mech 103(11):e202200600. https://doi.org/10.1002/zamm.202200600
8. Arif MS, Abodayeh K, Nawaz Y (2023) A computational scheme for stochastic Non-Newtonian mixed convection nanofluid flow over oscillatory sheet. Energies 16(5):5. https://doi.org/10.3390/en16052298
9. Nawaz Y, Arif MS, Abodayeh K, Mansoor M (2023) Finite difference schemes for MHD mixed convective Darcy–forchheimer flow of Non-Newtonian fluid over oscillatory sheet: a computational study. Front Phys 11. https://doi.org/10.3389/fphy.2023.1072296
10. Alqurashi MS et al (2023) Significance of melting heat in bioconvection flow of micropolar nanofluid over an oscillating surface. Sci Rep 13(1):11692. https://doi.org/10.1038/s41598-023-38361-w
11. Patel HR, Patel SD, Darji R (2022) Mathematical study of unsteady micropolar fluid flow due to non-linear stretched sheet in the presence of magnetic field. Int J Thermofluids 16:100232. https://doi.org/10.1016/j.ijft.2022.100232
12. Sharma B, Srivastava N (2023) Fluid flow past a rotating sphere in the presence of a toroidal magnetic field. ZAMM J Appl Math Mech Z Für Angew Math Mech 103(5):e202200218. https://doi.org/10.1002/zamm.202200218
13. Saini G et al (2023) Entropy generation optimization in couple stress fluid flow with variable viscosity and aligned magnetic field. Sustainability 15(3):3. https://doi.org/10.3390/su15032493
14. Maurya PK, Deo S, Maurya DK (2023) Couple stress fluid flow enclosing a solid sphere in a porous medium: effect of magnetic field. Phys Fluids 35(7):072006. https://doi.org/10.1063/5.0155532
15. Gogoi PP, Singh SK, Maity S (2023) Two layer thin film flow over a stretching sheet with uniform transverse magnetic field. J Magn Magn Mater 565:170204. https://doi.org/10.1016/j.jmmm.2022.170204
16. Dawar A et al (2020) Chemically reactive MHD micropolar nanofluid flow with velocity slips and variable heat source/sink. Sci Rep 10(1):20926. https://doi.org/10.1038/s41598-020-77615-9
17. Kumar D, Singh AK, Kumar D (2020) Influence of heat source/sink on MHD flow between vertical alternate conducting walls with hall effect. Phys Stat Mech Appl 544:123562. https://doi.org/10.1016/j.physa.2019.123562
18. Alqawasmi K et al (2023) Numerical approach toward ternary hybrid nanofluid flow with non-linear heat source-sink and fourier heat flux model passing through a disk. Int J Thermofluids 18:100367. https://doi.org/10.1016/j.ijft.2023.100367
19. Sharma RP, Badak K (2024) Heat transport of radiative ternary hybrid nanofluid over a convective stretching sheet with induced magnetic field and heat source/sink. J Therm Anal Calorim 149(9):3877–3889. https://doi.org/10.1007/s10973-024-12979-y
20. Manjunatha N, Yellamma, Sumithra R, Verma A, Gowda RJP, Madhu J (2024) The impact of the heat source/sink on triple component magneto-convection in superposed porous and fluid system. Mod Phys Lett B 38(07):2450020. https://doi.org/10.1142/S0217984924500209
21. Punith Gowda RJ, Naveen Kumar R, Jyothi AM, Prasannakumara BC, Sarris IE (2021) Impact of binary chemical reaction and activation energy on heat and mass transfer of Marangoni driven boundary layer flow of a Non-Newtonian nanofluid. Processes 9(4):4 (2021) https://doi.org/10.3390/pr9040702
22. Varun Kumar RS, Gunderi Dhananjaya P, Naveen Kumar R, Punith Gowda RJ, Prasannakumara BC (2022) Modeling and theoretical investigation on Casson nanofluid flow over a curved stretching surface with the influence of magnetic field and chemical reaction. Int J Comput Methods Eng Sci Mech 23(1):12–19. https://doi.org/10.1080/15502287.2021.1900451

23. Eric NM, Kinyanjui MN, Abonyo JO (2022) Two-phase turbulent fluid flow in a geothermal pipe with chemical reaction. J Appl Math 2022:e7617017. https://doi.org/10.1155/2022/7617017
24. Das UJ, Begum J (2023) Casson fluid flow with non-linear convective heat condition, Arrhenius activation energy and binary chemical reaction. AIP Conf Proc 2819(1):050013. https://doi.org/10.1063/5.0137104
25. Farooq U, Hussain M, Farooq U (2024) Non-similar analysis of chemically reactive bioconvective Casson nanofluid flow over an inclined stretching surface. ZAMM J Appl Math Mech Z Für Angew Math Mech 104(2):e202300128. https://doi.org/10.1002/zamm.202300128
26. Ali N, Ullah Khan S, Sajid M, Abbas Z (2016) MHD flow and heat transfer of couple stress fluid over an oscillatory stretching sheet with heat source/sink in porous medium. Alex Eng J 55(2):915–924. https://doi.org/10.1016/j.aej.2016.02.018
27. Al-Mubaddel FS, Allehiany FM, Nofal TA, Alam MM, Ali A, Asamoah JKK (2022) Rheological model for generalized energy and mass transfer through hybrid nanofluid flow comprised of magnetized cobalt ferrite nanoparticles. J Nanomater 2022:e7120982. https://doi.org/10.1155/2022/7120982
28. Sheikholeslami M, Hatami M, Ganji DD (2014) Nanofluid flow and heat transfer in a rotating system in the presence of a magnetic field. J Mol Liq 190:112–120. https://doi.org/10.1016/j.molliq.2013.11.002

Chapter 4
Characterization Methods for Nanomaterials in Energy Storage

B. C. Vasantha Kumar, R. Madhu Kumar, H. P. Shivarudrappa, B. M. Santhosh, and P. N. Prashantha Kumar

Abstract This chapter highlights the importance of characterization techniques in advancing energy storage materials and their role in shaping future energy technologies. Energy storage materials play a crucial role in advancing renewable energy technologies, enabling efficient energy capture, storage, and utilization. Characterization techniques are essential for understanding the physical, chemical, and electrochemical properties that govern their performance. Various characterization methods are employed in the study of energy storage materials, focusing on batteries, supercapacitors, and emerging storage technologies. Electrochemical techniques such as cyclic voltammetry and electrochemical impedance spectroscopy provide insights into charge storage mechanisms, ion diffusion rates, and electrode stability. Structural analysis using X-ray diffraction (XRD) and electron microscopy (SEM/TEM) reveals crystallographic structure, particle size, and morphology, influencing material stability, and cycling performance. Fourier-transform infrared spectroscopy (FTIR) identifies surface functional groups, oxidation states, and chemical composition changes during cycling. Understanding the complex interplay between structural, chemical, and electrochemical properties is paramount in optimizing energy storage materials for enhanced performance, durability, and sustainability. Continued advancements in characterization techniques promise to accelerate the development of next-generation energy storage solutions, facilitating the transition toward a cleaner and more efficient energy landscape.

Keywords Characterizations · Nanomaterials · Energy storage

B. C. Vasantha Kumar · P. N. Prashantha Kumar
Department of Chemistry, D.R.M. Science College, Davanagere, Karnataka 577004, India

R. Madhu Kumar
Department of Physics, RTE Society's Arts, Science, and Commerce Degree College, Ranebennur, Karnataka 581115, India

H. P. Shivarudrappa · B. M. Santhosh (✉)
Department of Chemistry, G.M. University, P.B. Road, Davanagere, Karnataka 577006, India
e-mail: santhoshamma041987@gmail.com

B. M. Santhosh
Visvesvaraya Technological University, Belagavi, India

R. Naik et al. (eds.), *Advanced Nanomaterials for Energy Storage Devices*, Nanostructure Science and Technology, https://doi.org/10.1007/978-3-031-74730-4_4

4.1 Introduction

Nanomaterials have the potential to influence a lot in scientific, commercial, and technological sectors because of their unique size-dependent properties [1]. However, substances lowered to the nanoscale possess distinct qualities that may deviate considerably from those of their bulk counterparts. Therefore, understanding and using the potential of nanomaterials requires accurate and comprehensive characterization techniques [2, 3]. The structural, chemical, physical, and optical properties of materials at the nanoscale are investigated and measured using a broad variety of analytical tools in nanomaterial characterization processes [4]. These techniques not only facilitate the construction and optimization of novel nanoscale structures for specific applications but they also aid scientists in comprehending the minute details of nanomaterials. These techniques not only facilitate the creation and optimization of novel nanoscale structures for specific applications but they also aid in better understanding of the minute details of nanomaterials by scientists. The minuscule size of nanomaterials, often varying from a few nanometers to several hundred nanometers, poses unique challenges for their characterization. Traditional bulk material characterization methods are sometimes inadequate at the nanoscale due to their low sensitivity and resolution.

At the nanoscale, conventional bulk material characterization methods are not always sufficient. The creation, evaluation, investigation, and utilization of nanostructured materials are the main foci of nanoscience. One billionth of a meter, or 10^{-9} m, is one nanometer (nm). A nanometer is roughly the length of 10 hydrogen atoms or 5 silicon atoms arranged in a line [5]. In recent years, there has been a great deal of interest in the processing, structure, and characteristics of materials whose grain sizes vary from tens to hundreds of nanometers. The field of materials science and engineering is undergoing a revolution as scientists discover methods to design and describe materials with a length scale of nanometers. Innovative materials possessing exceptional electrical, optical, magnetic, and mechanical characteristics are being created at a rapid pace for applications in bioengineering, information technology, energy, and the environment. Some physical and chemical properties of materials can differ significantly at the nanoscale from those of bulk structured materials of the same composition; for instance, the theoretical strength of nanomaterials can be reached or quantum effects may appear; crystals in the nanometer scale have reduced lattice constants and a low melting point (the difference can be as large as 1000 °C) because the number of surface atoms or ions becomes a significant fraction of the total number of atoms or ions and the surface energy plays a significant role in the thermal stability [6].

The creation of affordable, scalable, dependable, and efficient catalysts is crucial for achieving this goal. It has been demonstrated that carbon-based materials have a great deal of potential for enhancing the efficiency and dependability of energy conversion and storage systems [7]. In the realm of energy storage, a variety of nanomaterials have come into focus following decades of study and development. Carbon materials are thought to serve a crucial function as electrode materials, conductive agents, etc., in addition to the materialization of other materials in energy storage.

Nanomaterials are engineered or chosen for their nanoscale attributes because these attributes offer new or enhanced functionalities. In defining "nano" for characterization purposes, lengths are typically categorized as macro (> mm), micro (mm–μm), nano (μm–nm), and atomic (less than nm). Nanomaterials are pivotal in the field of energy storage due to their distinct properties, including a high surface area, improved conductivity, and better electrochemical performance. While standard macroscale characterization is often sufficient for the development of energy storage systems, it is crucial to characterize these materials to comprehend their structure, composition, and properties, which significantly affect their performance in energy storage systems.

4.2 Experimental Details: Theory and Specifications

Detailed information on the various characterization techniques employed to investigate the physical, chemical, thermal, morphological, magnetic, and shielding properties is given here.

4.2.1 X-Ray Diffraction Study

To determine the solid-state properties of the samples, X-ray diffraction experiments were conducted. The samples underwent scanning within the 2θ range from 10 to 60°, with a scanning rate of 1°/ml and a step size of 0.02°. The target voltage for operation was set at 40 kV, and the tube current was maintained at 100 mA. The X-ray source used was CuKα radiation, with a wavelength of 1.5406 Å. Based on the width of the diffraction patterns seen in the X-ray reflected crystalline area, the average crystallite size of the crystallites was determined using the technique and the Scherrer equation. The diffraction pattern that was acquired from the lattice planes at the diffraction peak was used in this investigation to calculate the crystallite size [8],

$$L_{XRD} = \frac{K\lambda}{\beta \cos\theta} \tag{4.1}$$

where diffraction angle is α, whole breadth at half maximum value of the observed reflection is β, incoming X-ray wavelength is 1.5406 Å, and k is the Scherer constant (0.94).

The tangent formula may be used to compute the average lattice strain (ε_{av}), which is the average strain caused by the lattice deformation:

$$\varepsilon_{\text{av}} = \frac{\beta}{4\tan\theta} \tag{4.2}$$

Crystalline peaks are placed atop an amorphous halo in polycrystalline materials. This implies that the crystallite formations, which are composed of normal polymer chains, are likely to be amorphous because to the uneven chain arrangement in their sea. The following formula can be used to determine the average spacing between these crystallites [9].

$$R = \frac{5\lambda}{8\sin\theta} \tag{4.3}$$

The level of structural organization within the sample, represented by crystallinity, can also be determined based on the proportions of the regions below the crystalline main line and the accompanying halos, according to Herman's and Weidinger's technique [10]:

$$X_C(\%) = \frac{A_c}{A_c + A_a} \times 100 \tag{4.4}$$

where A_c and A_a are the areas of crystalline and amorphous (halo) regions in the crystalline peak, respectively.

4.2.2 *Fourier Transform Infrared Spectroscopy*

The various chemical bonds contained in a sample can be seen using the non-destructive Fourier transform infrared (FT-IR) spectroscopy method. The organic and inorganic compounds contained in the sample can be identified with the aid of the FT-IR. The different chemical bonds in the sample will absorb the IR radiation's characteristic band of frequencies and begin to vibrate and rotate when it is incident upon.

The presence of a certain chemical bond in the sample will then be confirmed by a number of transmission peaks that we will see in the wavenumber spectrum. The quantity of bonds or components in the sample will also be disclosed by the peak's magnitude. Even while it will not provide the sample's entire chemical structure, it will enable us to see the changes the sample has undergone chemically. As a result, FT-IR is a particularly useful tool for observing changes in the transmission or absorption peaks caused by the sample's exposure to different impacts, such as blending, aging, heat treatment, chemical treatment, and filler addition [11]. Prior to scanning the real sample, a blank scan is performed in the FT-IR spectrometer to avoid background transmission peaks. The result of this scan is then deducted from the sample data.

The transmittance is calculated as follows:

$$T\% = \frac{\text{Intensity of the sample } I_t}{\text{Intensity of reference beam } I_o} \times 100 \tag{4.5}$$

4.2.3 Scanning Electron Microscopy

One of the most well-liked and often applied methods for the characterizations of nanomaterials and nanostructures is scanning electron microscopy (SEM). SEM is an efficient tool for characterizing specimens at a resolution of a few nanometers, with image magnifications of between 10 and 30,000 times are possible. SEM can offer helpful information on chemistry, crystal orientation, and internal stress distribution in addition to surface topography. Electrons are focused into a beam with a tiny 5 nm spot size, using electron cannon in SEM. Electrons are accelerated by deflection coils to energies of a few hundred eV – 50 keV, after which they are restored across the surface of the specimen [12]. By collecting the liberated electrons as they hit and penetrate the surface on a cathode ray tube (CRT), SEM images are produced. One can distinguish between different SEM methods based on what is later detected and photographed. SEM produces three main image types: backscattered electron images, secondary electron images, and elemental X-ray maps.

When a high-energy primary electron interacts with an atom, two processes take place: inelastic scattering with atomic electrons and elastic scattering with the atomic nucleus. In an inelastic collision involving five electrons, the initial electron imparts some of its energy to the secondary electron. When the energy transmitted is great enough, the other electron will come free from the sample. If the energy of an electron emitted from one of the five orbitals of the incident atom is less than 50 eV, it is called a secondary electron. Since the secondary electron energy is small, the SEM images are highly sensitive to topographic variations. *Backscattered electrons* are high-energy incident electrons that are elastically scattered and essentially possesses nearly the same energy as the incident or primary electrons.

An additional electron interaction occurs when the main electron in a SEM collides with and ejects a core electron from an atom in the sample. When the excited atom decays to its ground state, two useful particles for chemical characterization are released: an Auger electron or an X-ray photon.

The theoretical limit of the resolving power of an instrument is determined by the numerical aperture around the system and the wavelength of the electron beam being used. The resolving power, or R, of an instrument is defined as follows:

$$R = \frac{\lambda}{2\text{NA}} \tag{4.6}$$

where NA is the numerical aperture, etched on each objective and condenser lens system, and λ is the wavelength of electrons used. It also serves as a measure of the objective's or the condenser's ability to provide electrons.

4.2.4 Transmission Electron Microscopy

Transmission electron microscopy (TEM) is a vital tool in the field of nanotechnology for investigating the imaging of nanomaterials at sub-nanometer resolution. Albert Prebus and James Hillier at the University of Toronto in 1938 used Max Knoll and Ernst Ruska's concept to create the first practical TEM. Using this technique, an electron beam with a constant current density is used to irradiate a thin material [13]. The standard instrument's acceleration voltage ranges from 80 to 120 kV. The medium-voltage instrument works with an acceleration voltage between 200 and 500 kV to enhance transmission and resolution. For high-voltage electron microscopy, the acceleration voltage range is 500 kV–3 MV. Through Schottky, field, or thermionic emission (tungsten or lanthanum hexaboride filaments), the electrons in the electron cannon are released. Both the illumination aperture and the illuminated specimen's area are controlled by the two-stage condenser lens. The objective lens's main job is to take a picture or generate a diffraction pattern on the specimen. These electron diffraction patterns provide information about the crystallographic structure of the sample. A photo may be taken by the photographic plate of the digital fluorescence screen, which is linked to a CCD camera through a fiber-optic plate. The goal of adding more state-of-the-art methods to the apparatus is to boost TEM analysis's capacity for example, dispersive X-ray analysis (EDX) and electron energy loss spectroscopy (EELS).

4.2.5 Raman Spectroscopy

Raman spectroscopy is a powerful and versatile method when evaluating materials with nonelastic light scattering of continuously vibrating intramolecular links [14]. Every molecule achieves a "virtual state" when stimulated by a high-power laser light, and it immediately releases energy by a variety of processes, such as nonelastic scattering [15]. The inelegant change in frequency caused by this light scattering is called Raman scattering, and it may be recognized and analyzed to provide information on the chemical composition of the material.

Raman spectra are acquired during the process, which includes the "molecular fingerprint" of the sample under examination and can reveal crucial information about its composition and structure [16, 17]. A common application of Raman spectroscopy is in biomedicine. Examining the composition of tissues and cells, for example, can provide important information about the health and functionality of those cells. It can also be used to study the structure and conformation of proteins, which is essential to understanding how they function, or to look into how drugs interact with biomolecules.

All things considered, Raman spectroscopy is a useful tool in the field of biomedicine because it can give precise information about the molecular makeup of biological samples. Raman spectroscopy can be used to evaluate naturally occurring polysaccharide polymers by identifying functional groups and, consequently, their chemical composition. This allows for an understanding of the polymers' chemical and physical properties.

Several methods are employed in the instrumental analysis of polymers: Raman spectroscopic imaging techniques: coherent anti-Stokes Raman scattering (CARS), surface-enhanced Raman spectroscopy (SERS), tip-enhanced Raman spectroscopy (TERS), and microspectroscopy (RSI).

The Raman spectrometer uses a monochromator, detector, sample holder, and laser as its light source. These days, lasers are used as excitation light sources for Raman spectroscopy. Depending on how specialized the study is, different types of lasers will be used. The choice of laser excitation wavelength is one of the most important aspects of Raman spectroscopy. Any of the laser types that emit light at a specific wavelength can serve as the excitation source, including diode lasers (630 and 780 nm), argon (488.0 and 514.5 nm), krypton (530.9 and 647.1 nm), helium–neon (632.8 nm), neodymium (1064 and 532 nm), and krypton (530.9 and 647.1 nm).

The fluorescence effect is less when a near-infrared (NIR) excitation laser with a wavelength of 1064 nm is used as the light source than when visible wavelength lasers are used [18]. The choice of wavelengths is typically made by weighing the trade-offs between maximizing the intensity of the Raman signal and minimizing background autofluorescence, both of which increase with decreasing wavelength (the excitation frequency of the Raman signal scales to the fourth power). The type of sample and application have a significant impact on how well this balance is optimized. When stimulated with a NIR laser at 785–1048 nm, the majority of tissues produce a sufficient Raman signal and the intended low autofluorescence [19]. There are, however, some exceptions, as some tissues may exhibit excessive autofluorescence even in response to NIR excitation [20]. Shorter excitation wavelengths, like 633 nm or 532 nm, are frequently advantageous for samples with exceptionally low rates of autofluorescence because they generate stronger signals and drastically reduce acquisition time. However, it should be mentioned that because of the potential effects of molecular resonance, Raman spectra acquired at various laser excitation wavelengths shouldn't be directly compared.

A system of mirrors and lenses focuses the excitation radiation onto the test sample, scattering it there. Raman microscopy imaging resolution varies depending on the type of system, but it is ideally limited by diffraction, which means that the choice of lens and excitation wavelength are crucial [21].

4.2.6 Electrochemical Impedance Spectroscopy (EIS)

The measurement of electrical impedance in the widest possible frequency range is the basis of a number of characterization techniques of solids, liquids, and suspensions of biological tissues and fluids in vitro of semiconductor devices it is employed in the testing of electronic components, of engineering materials and structures, and for sensor reading and conditioning. The result of such impedance spectroscopy measurements is a set of complex (magnitude and phase) spectra of impedance versus frequency. The spectra typically have smooth and small features; often, a few percent variation in magnitude, less than a degree variation in phase angle or a small change in the frequency defining the peak of a broad feature, represent all the experimental information available. Therefore, high measurement accuracy, and strict metrological control of the instrument employed, become mandatory to achieve meaningful measurements.

Nowadays, commercial instruments are available having frequency bandwidths extending from a few Hz to tens of MHz. Those instruments are easy to operate and have a fast response. Resolutions of six or seven digits, and high measurement repeatability (often better than 10^{-5}) are not uncommon in top-class instruments; such precision gives the operator the false feeling of a corresponding very high accuracy. In fact, relative accuracy specifications can be in the 10^{-4} range at audio frequencies but can degrade to a few percent beyond 100 kHz [22].

In the AC Impedance technique (IMP), the base potential is held constant at Init *E*. A sine waveform is superimposed onto the base potential, and its frequency is scanned from high to low with 12 components per decade. Current and potential are sampled and analyzed to obtain the real and imaginary parts of the impedance [23].

Electrical impedance, as a physical quantity, can be defined at any frequency f. Impedance metrology refers, however, to two very different scientific fields. In electrical metrology, the impedance Z is defined as the complex ratio of the phasor voltage V between specified given terminals of a device and the phasor current I flowing through a given terminal of the same device. Typical measurement frequency ranges from 50 Hz to 20 kHz, with a particular interest for $f = 1000$ Hz or 1592 Hz (the latter corresponds to $\omega = 10\ \text{krad s}^{-1}$). In radio-frequency electromagnetic metrology, impedance Z is defined at a section of a device (typically identified with a coaxial port) and is a quantity related to the scattering of an electromagnetic wave incident on the section, with reference to the characteristic impedance of the measurement system. Typical measurement frequency ranges from 50 MHz to hundreds of GHz [24].

4.3 Characterization Methods for Nano-Materials in Energy Storage Applications

Characterization methods for nanomaterials in energy storage applications are crucial for understanding their structure, properties, and performance. Here are some common characterization methods for nanomaterials in energy storage:

4.3.1 X-Ray Diffraction (XRD)

XRD is used to determine the crystal structure and phase composition of nano-materials. By analyzing the diffraction pattern produced when X-rays interact with the sample, researchers can identify the crystalline phases present and measure parameters such as crystallite size and strain. X-ray Diffraction (XRD) is a vital characterization technique for nanomaterials in energy storage applications.

XRD is commonly employed to identify the crystalline phases present in nanomaterials used in energy storage devices. Many electrode materials, such as transition metal oxides, sulfides, and phosphates, exhibit polymorphism or can form multiple phases during synthesis or electrochemical cycling. XRD provides insights into the crystal structure and lattice parameters of nano-materials, facilitating the characterization of their atomic arrangement and symmetry. By analyzing the positions and intensities of diffraction peaks, researchers can derive structural parameters such as unit cell dimensions, crystallite size, and lattice strain. The broadening of diffraction peaks in XRD patterns is indicative of smaller crystallite sizes or lattice distortions induced by defects, strain, or dislocations [25]. Smaller crystallite sizes can enhance ion diffusion kinetics and surface reactivity, while controlled strain levels can mitigate mechanical degradation and enhance structural stability during charge–discharge cycles [26]. XRD can assess the phase purity and degree of crystallinity of nano-materials, as well as detect the presence of impurities or secondary phases that may affect their electrochemical properties. Energy storage materials synthesized via different methods or subjected to various processing conditions may exhibit variations in phase composition and purity. Advanced XRD techniques, such as in situ and operando XRD, enable real-time monitoring of structural changes in nanomaterials during electrochemical processes [27]. These techniques allow researchers to track phase evolution, lattice dynamics, and stress-induced transformations in electrodes under operating conditions, providing dynamic insights into mechanisms such as phase transitions, lattice expansion, and amorphization. By correlating XRD data with electrochemical performance metrics, the structure–property relationships can be elucidated and guide the design of nanomaterials with improved energy storage performance and durability.

4.3.2 Fourier Transform Infrared Spectroscopy (FTIR)

FTIR is employed to analyze the chemical composition and functional groups present in nano-materials. It provides information about bonding characteristics, surface chemistry, and the presence of impurities or contaminants that may affect the performance of energy storage devices. Understanding the chemical composition of electrode materials is crucial for optimizing their electrochemical properties, stability, and compatibility with electrolytes in energy storage devices. FTIR provides information about the chemical composition and functional groups present in nanomaterials used in energy storage devices. By analyzing the absorption bands in the infrared spectrum, the molecular vibrations associated with specific chemical bonds, such as O–H, C–H, C=O, and metal–oxygen bonds in metal oxides can be identified. This enables qualitative and quantitative analysis of materials, including the detection of impurities, surface species, and functional coatings. FTIR is sensitive to surface functional groups and modifications in nano-materials, which play a significant role in electrode–electrolyte interactions and charge transfer processes in energy storage devices [28]. Surface functionalization techniques, such as doping, surface coating, or chemical modification, can alter the surface chemistry and properties of nano-materials, affecting their electrochemical behavior and stability. FTIR analysis enables to characterize surface modifications, assess their impact on surface reactivity and wettability, and optimize electrode designs for enhanced energy storage performance and cycling stability [29].

FTIR can be used for quantitative analysis of surface species, such as adsorbed species, functional groups, and surface contaminants, which influence the surface reactivity and electrochemical activity of nano-materials. By comparing the intensity of specific absorption bands with reference standards or calibration curves, it is possible to determine the concentration or coverage of surface species on electrode materials. This information is valuable for understanding surface reactions, ion adsorption–desorption processes, and the formation of passivation layers (e.g., solid-electrolyte interfaces) in energy storage devices [30].

FTIR can be coupled with electrochemical techniques to study in situ or operando electrochemical processes occurring at electrode–electrolyte interfaces in energy storage devices. Techniques such as attenuated total reflection (ATR)-FTIR or infrared spectroelectrochemistry enable real-time monitoring of changes in molecular species, functional groups, and chemical bonds during electrochemical cycling. This provides insights into electrochemical reactions, electrolyte decomposition, and the formation of surface films, guiding the optimization of electrode materials and electrolyte formulations for improved energy storage performance and durability. FTIR is sensitive to surface adsorption phenomena and interactions between nanomaterials and electrolyte ions in energy storage devices. The infrared spectra of materials taken before and after exposure to electrolytes led to investigate ion adsorption–desorption processes, ion coordination environments, and the formation of ion-conducting pathways in electrodes. Understanding surface-ion interactions is essential for optimizing electrode–electrolyte interfaces, enhancing ion diffusion kinetics, and mitigating degradation mechanisms such as electrolyte decomposition and dendrite formation in energy storage devices [31].

4.3.3 Scanning Electron Microscopy (SEM)

Scanning Electron Microscopy (SEM) is a powerful technique for characterizing nanomaterials in energy storage applications. SEM provides high-resolution images of the surface morphology of nano-materials. The main electron beam penetrates the sample and scatters and absorbs in the "interaction volume," a droplet-shaped area with a typical size of ~ 1 μm. Specialized detectors can detect electron elastic scattering, secondary electron emission by inelastic scattering, and X-ray photon emission resulting from the interaction of the source beam with the material. Because beam scanning is timed to coincide with an image projection, the final picture is a distribution of the signal's intensity that is released from the sample. Since SEM is a scanning method, it technically lacks magnification. Instead, the beam size and precise placement on the sample determine the field of view and final picture. The ratio of the scanned sample area to the displayed picture size may be used to calculate the magnification. It allows researchers to visualize the size, shape, and distribution of nanoparticles, as well as any structural defects or agglomerations present in the material.

SEM is used in battery research to study particle morphology and size distribution, investigation of films or coatings formation, uniformity and smoothness, dendrite formation, electrode microstructure, impurities, ion diffusion, and contamination as well [32]. SEM provides high-resolution images of the surface morphology of nanomaterials. For energy storage devices like batteries and supercapacitors, the surface structure of electrodes significantly influences their performance. SEM technique is used to visualize the size, shape, and distribution of nanoparticles, as well as any structural defects or agglomerations present in the material. This information helps in understanding how the morphology affects the material's electrochemical behavior, such as ion diffusion and charge storage capacity.

Vigorous research and development in the field of Li-ion batteries necessitates high energy density products that are safe and reliable. Metallic lithium is the most promising and dense anode [33]. However, the problem of dendritic development, which affects anode materials, is more noticeable in materials that include elemental metal. The dendrite development can puncture an ion-transporting separator between the cathode and anode; it poses a risk to battery safety [34]. In this context, SEM has been used extensively in research on the dendritic growth issue. There have been reports of several methods to prevent dendrite development, including coatings, 3D patterning of Li anode surfaces, and generating a constant Li^+ flow over the surface using specialized separators [35]. SEM analysis of the electrode and separator morphology and composition is necessary for all of these methods.

SEM can accurately measure the size and distribution of these nanoparticles, which is crucial for optimizing the electrochemical performance of energy storage devices. Uniform particle size distribution and controlled morphology can enhance electrode packing density, increase surface area, and improve electrode–electrolyte interaction, leading to better energy storage capacity and cycling stability [36]. The electrical performance of a material is also greatly influenced by its shape.

Particle size affects material intercalation performance, Li^+ diffusion rate, and electric conductivity [37]. Micron-sized particles have greater long-term stability with lower C-rate capacity, whereas nanosized materials often show outstanding high C-rate capability but may have substandard long-term stability [38].

In energy storage devices, the interface between electrode materials and electrolytes plays a vital role in determining their performance. SEM enables us to study the electrode–electrolyte interface at high resolution, providing insights into factors such as interfacial contact, wetting behavior, and the formation of solid–electrolyte interfaces (SEI). Understanding these interfaces is essential for optimizing ion transport kinetics, reducing interfacial resistance, and mitigating degradation mechanisms, thereby improving device efficiency and lifespan. In situ SEM microscopy is commonly employed in battery research to examine the development of SEI and coating formation. An ionic liquid electrolyte, which does not evaporate even at extremely high vacuums, is necessary for a standard in situ SEM cell probe [39]. Since the porosity, morphology, and microstructure of the electrode materials impact the penetration rate and availability of oxygen molecules, which are reversibly converted in the battery, SEM is highly desired for research on metal–oxygen batteries (e.g., $Li–O_2$, $Zn–O_2$) [40].

Advanced SEM techniques, such as in situ and operando microscopy, allow real-time observations of nanomaterials during electrochemical cycling. These techniques is used to monitor structural and morphological changes in electrodes under operating conditions, providing dynamic insights into phenomena such as particle fracturing, dendrite formation, and electrode–electrolyte interactions. By correlating SEM observations with electrochemical performance data, researchers can optimize electrode design and develop strategies to mitigate degradation pathways, advancing the development of high-performance energy storage devices.

4.3.4 Transmission Electron Microscopy (TEM)

The transmission electron microscope (TEM) operates on the same principles as an optical one, but it may achieve much greater magnification due to the tiny de Broglie wavelength of electrons. On the other hand, distinct aberration processes restrict the resolution of the TEM. With the advent of chromatic aberration correctors, TEM resolution has recently increased significantly, allowing the user to study minute features like individual molecules, atoms, or vacancies. The majority of TEMs are also capable of operating in the scanning transmission electron microscopy (STEM) mode. In this mode, the beam scans over the sample while being focused to a narrow diameter and the signal that is emitted is recorded. This mode allows for chemical element mapping using EELS or EDS.

Transmission Electron Microscopy (TEM) offers even higher resolution than SEM and enables the detailed examination of nanostructures at the atomic level. It provides information about the crystalline structure, lattice spacing, and grain boundaries of energy storage materials such as cathodes and to study the thin layer coatings that are placed to the surface of electrode materials. Thin coatings are important because they promote electronic conductivity and decrease capacity loss with cycling on electrode material surfaces. TEM is incredibly helpful to investigate local phase transitions and reactions that take place on the coating interface and comprehend how they affect the final material's characteristics [41]. In situ TEM investigations of phase transitions during Li intercalation, including imaging, ED, EDS, and EELS measurements, can show phase and morphological changes, chemical reactions throughout the charge–discharge process, and help identify reaction products [42].

The creation of high-performance electrode materials necessitates an understanding of the homogeneity of the samples and active or reactive sites during various charge–discharge phases. TEM can be used to investigate the locations of such sites. Li atoms diffuse through the walls of carbon nanotubes, resulting in bead-like formations, despite the expectation that Li should enter the carbon nanotubes from their ends, as demonstrated by in situ TEM examinations [43]. Li intercalation into carbon nanotubes is a totally reversible process, as evidenced by the same morphologies and diameter distributions of the samples before and after intercalation [44].

With an in situ TEM, the kinetics of the processes taking place in electrochemical cells may also be investigated. The SEI is the focus of many in situ investigations; its production, location, and change over time are all important factors in the development of novel electrode materials, like silicon. SEI generation may occur on the surface of material coatings due to a unique electrode material design but not on the active phase itself. An illustration of a material resembling pomegranates like structure is silicon wrapped into carbon spheres [45]. Such materials' in situ TEM investigations shed light on the lithiation kinetics, spare volume retained inside the carbon shell, and SEI generation. In situ TEM investigations of these kinds of materials reveal details on the development of SEI, the kinetics of lithiation, and the spare volume remaining inside the carbon shell. Electrochemical cells with graphene windows, for instance, allow researchers to detect individual atoms and the atomic structure of individual particles in situ during electrochemical charge–discharge processes, with liquid electrolyte flowing through the cell. This is an example of how TEM techniques are being further developed.

4.3.5 Raman Spectroscopy

Raman spectroscopy is used to study the vibrational modes of molecules in nanomaterials. It can provide insights into the molecular structure, symmetry, and electronic properties of materials, as well as detect defects, doping, and phase transitions. Raman spectroscopy is a powerful characterization technique for nanomaterials in energy storage applications.

Raman spectroscopy provides information about the molecular vibrations and phonon modes present in nano-materials. Each material has a unique Raman spectrum, which consists of peaks corresponding to specific vibrational modes of chemical bonds and lattice vibrations. This enables qualitative and quantitative analysis of materials, including the detection of impurities, phase transitions, and surface modifications [46]. Raman spectroscopy offers insights into the structural properties and defects of nanomaterials at the microscopic level. Raman scattering is sensitive to changes in crystal symmetry, lattice strain, and disorder, helps to investigate structural features such as crystallinity, grain boundaries, and defects in electrode materials [47]. This information is crucial for understanding the relationship between material structure and electrochemical performance, including ion diffusion kinetics, charge storage mechanisms, and mechanical stability in energy storage devices. Many energy storage materials, such as transition metal oxides, sulfides, and phosphates, exhibit polymorphism or can form multiple phases during synthesis or electrochemical cycling [48].

Raman spectroscopy is sensitive to surface phenomena and interactions between nanomaterials and electrolytes in energy storage devices. By studying Raman spectra before and after exposure to electrolytes, one can investigate surface adsorption–desorption processes, formation of passivation layers (e.g., solid-electrolyte interfaces), and chemical reactions occurring at electrode–electrolyte interfaces [49]. Understanding surface and interface chemistry is essential for optimizing ion transport kinetics, enhancing electrochemical stability, and mitigating degradation mechanisms in energy storage devices. Advanced Raman spectroscopy techniques, such as in situ and operando Raman spectroscopy, enable real-time monitoring of structural and chemical changes in nanomaterials during electrochemical processes.

4.3.6 BET Surface Area Analysis

Brunauer–Emmett–Teller (BET) analysis is used to measure the specific surface area of nano-materials, which is crucial for understanding their adsorption capacity, ion diffusion kinetics, and electrochemical activity in energy storage applications. BET surface area analysis is a crucial technique for characterizing nanomaterials in energy storage applications.

BET analysis quantifies the specific surface area of nanomaterials, which is a critical parameter influencing their performance in energy storage devices. Nanomaterials typically have high surface area-to-volume ratios due to their small particle size and high porosity. By measuring the surface area using gas adsorption techniques, researchers can assess the available surface area for electrochemical reactions, ion adsorption–desorption processes, and electrolyte interaction. Higher surface areas generally correlate with enhanced electrode–electrolyte contact, improved ion diffusion kinetics, and increased electrochemical activity, leading to higher energy storage capacity and efficiency [50]. BET analysis provides information about the pore size distribution and pore volume of nanomaterials, which are crucial factors affecting their ion storage and transport properties. Nanomaterials often contain a variety of pores, including micropores, mesopores, and macropores, with different sizes and geometries. Understanding the pore size distribution allows investigators to optimize the accessibility of electrolyte ions to active sites within the material, maximize ion diffusion rates, and minimize diffusion limitations. This is particularly important for energy storage devices such as supercapacitors and lithium-ion batteries, where fast charging and discharging rates are desired.

BET analysis provides insights into the surface morphology and porosity of nanomaterials, including features such as surface roughness, particle aggregation, and hierarchical pore structures. Surface morphology and porosity significantly influence the effective surface area available for electrochemical reactions and ion transport processes. BET analysis involves plotting the adsorption isotherm of gas molecules (typically nitrogen) on the surface of nanomaterials. The shape of the adsorption isotherm provides information about the surface properties and pore structure of the material. For example, a type IV isotherm with a hysteresis loop indicates the presence of mesopores, while a type I isotherm suggests the predominance of micropores [51]. BET analysis can assess the impact of surface functionalization and modification techniques on the surface area and porosity of nanomaterials. Surface functionalization involves introducing functional groups or coatings to modify the surface chemistry and properties of nanomaterials, influencing their electrochemical activity and stability [52].

4.3.7 Electrochemical Impedance Spectroscopy (EIS)

EIS is a powerful technique for studying the electrochemical behavior of nanomaterials in energy storage devices. It provides information about charge transfer kinetics, ion diffusion coefficients, and the resistance of interfaces and electrolytes, helping to optimize device performance.

EIS provides valuable insights into the electrochemical behavior of nanomaterials and electrodes in energy storage devices. By measuring the impedance response of the system over a range of frequencies, EIS can extract information about charge transfer kinetics, ion diffusion rates, and interfacial processes occurring at electrode–electrolyte interfaces [53]. This enables researchers to understand the mechanisms governing energy storage performance, such as electrode kinetics, double-layer capacitance, and diffusion-controlled processes, helping to optimize device design and operation [54]. EIS allows for the characterization of the electrode–electrolyte interface, which is crucial for determining the overall performance and stability of energy storage devices. Understanding and optimizing the interface properties are essential for improving device efficiency, cycling stability, and safety. EIS can elucidate degradation mechanisms and failure modes in energy storage devices, providing early warning signs of performance degradation and aging. Changes in impedance spectra over time or cycling can indicate phenomena such as electrode dissolution, electrolyte depletion, electrode–electrolyte interface evolution, and formation of irreversible reaction products [55]. EIS is also used to characterize the electrochemical properties of nanomaterials, such as conductivity, diffusivity, and electrochemical activity, which are critical for energy storage applications. This information aids in the selection and optimization of nanomaterials for specific energy storage applications, guiding material synthesis, processing, and integration into devices.

4.3.8 Cyclic Voltammetry (CV)

CV is commonly used to characterize the electrochemical behavior and stability of nanomaterials. By measuring the current response to applied voltage cycling, researchers can evaluate the capacitance, redox activity, and cycling stability of electrode materials for energy storage applications.

CV provides insights into the electrochemical behavior of nanomaterials and electrodes in energy storage devices. By measuring the current response to a linearly varying potential sweep, CV generates cyclic voltammograms that reveal information about redox reactions, charge storage mechanisms, and electrochemical kinetics occurring at electrode–electrolyte interfaces. CV allows for the characterization of redox-active nanomaterials and their charge storage mechanisms in energy storage devices. Peaks in the cyclic voltammogram correspond to oxidation and reduction reactions associated with specific redox couples or electroactive species in the electrode material. The shape, position, and intensity of these peaks provide information about the electrochemical reversibility, redox potential, and capacity of the material. Additionally, CV can distinguish between Faradaic processes (e.g., intercalation/deintercalation) and non-Faradaic processes (e.g., double-layer capacitance), enabling researchers to assess the contribution of each mechanism to overall charge storage and ion diffusion kinetics, aiding in the optimization of device design and material selection [54].

CV analysis allows for the determination of electrode kinetics and reaction rates in energy storage materials. By analyzing the shape and slope of the voltammograms, parameters such as peak potentials, peak currents, and peak separation, which are related to electron transfer kinetics, charge transfer resistance, and reaction rates at electrode surfaces, can be studied [56]. Understanding the kinetics of electrochemical processes is crucial for optimizing electrode design, material morphology, and operating conditions to achieve fast charge–discharge rates, high power density, and efficient energy storage performance. CV can assess the compatibility and stability of nanomaterials with electrolytes used in energy storage devices. Changes in the CV response, such as shifts in peak potentials or variations in peak shapes, can indicate interactions between the electrode material and electrolyte, electrolyte decomposition, or formation of passivation layers. Understanding electrolyte-material interactions is essential for selecting suitable electrolytes and optimizing device performance and lifespan.

4.4 Conclusion

In summary, X-ray diffraction is a powerful tool for characterizing nanomaterials in energy storage applications, offering valuable information about phase composition, crystal structure, crystallite size, and strain. Fourier transform infrared spectroscopy is a versatile tool for characterizing nanomaterials in energy storage applications, providing valuable insights into their chemical composition, surface chemistry, electrochemical processes, and interactions with electrolytes. SEM and TEM are the versatile tools for characterizing nanomaterials in energy storage applications, providing valuable information about surface morphology, particle size, interface properties, structural defects, and degradation mechanisms. Raman Spectroscopy is a versatile tool for characterizing nanomaterials in energy storage applications, providing valuable information about their molecular structure, crystallographic properties, surface chemistry, and electrochemical behavior. BET surface area is used for characterizing nanomaterials in energy storage applications, providing essential information about surface area, pore size distribution, surface morphology, and porosity. Electrochemical impedance spectroscopy is a versatile technique for characterizing nanomaterials in energy storage applications, providing valuable information about electrochemical behavior, interface properties, degradation mechanisms, material properties, and device performance. CV is a versatile technique for characterizing nanomaterials in energy storage applications, providing valuable information about electrochemical behavior, redox activity, kinetics, electrolyte compatibility, and material performance. By employing these characterization techniques, researchers can gain valuable insights into the structure–property relationships of nanomaterials and tailor their design for next-generation batteries, supercapacitors, and other energy storage technologies.

References

1. Khan Y, Sadia H, Ali Shah SZ, Khan MN, Shah AA, Ullah N, Khan MI (2022) Classification, synthetic, and characterization approaches to nanoparticles, and their applications in various fields of nanotechnology: a review. Catalysts 12(11):1386. https://doi.org/10.3390/catal12111386
2. Liu X, Meng J, Zhu J, Huang M, Wen B, Guo R, Mai L (2021) Comprehensive understandings into complete reconstruction of precatalysts: synthesis, applications, and characterizations. Adv Mater 33(32):2007344. https://doi.org/10.1002/adma.202007344
3. Tambo F, Ablateye DNO (2020) A review on the role of emerging revolutionary nanotechnology in forensic investigations. J Appl Natural Sci 12(4):582–591. https://doi.org/10.31018/jans.v12i4.2415
4. Salem SS, Hammad EN, Mohamed AA, El-Dougdoug W (2022) A comprehensive review of nanomaterials: types, synthesis, characterization, and applications. Biointerface Res Appl Chem 13(1):41. https://doi.org/10.33263/BRIAC131.041
5. Jayawardena HSN, Liyanage SH, Rathnayake K, Patel U, Yan M (2021) Analytical methods for characterization of nanomaterial surfaces. Anal Chem 93(4):1889–1911. https://doi.org/10.1021/acs.analchem.0c05208

6. Joudeh N, Linke D (2022) Nanoparticle classification, physicochemical properties, characterization, and applications: a comprehensive review for biologists. J Nanobiotechnol 20:262. https://doi.org/10.1186/s12951-022-01477-8
7. Kaygusuz K (2012) Energy for sustainable development: a case of developing countries. Renew Sustain Energy Rev 16(2):1116–1126. https://doi.org/10.1016/j.rser.2011.11.013
8. Torrado G, Fraile S, Torrado S, Torrado S (1998) Process-induced crystallite size and dissolution changes elucidated by a variety of analytical methods. Int J Pharmaceutics 166(1):55–63. https://doi.org/10.1016/S0378-5173(98)00021-0
9. Ali TM, Padmanathan N, Selladurai S (2015) Effect of nanofiller CeO_2 on structural, conductivity, and dielectric behaviors of plasticized blend nanocomposite polymer electrolyte. Ionics 21:829–840. https://doi.org/10.1007/s11581-014-1240-y
10. Hermans PH, Weidinger A (1961) On the determination of the crystalline fraction of polyethylenes from X-ray diffraction. Macomole Chem Phys 44(1):24–36. https://doi.org/10.1002/macp.1961.020440103
11. Tazibt N, Kaci M, Dehouche N, Ragoubi M, Atanase LI (2023) Effect of filler content on the morphology and physical properties of poly(lactic acid)-hydroxyapatite composites. Materials 16(2):809. https://doi.org/10.3390/ma16020809
12. Ul-Hamid A, Ul-Hamid A (2018) Imaging with the SEM. In: A beginners' guide to scanning electron microscopy, 129–180. https://doi.org/10.1007/978-3-319-98482-7_9
13. Reimer L, Kohl H (2008) Physics of image formation. Transmission electron microscopy, 4th edn. Springer-Verlag, Berlin, New York, 36
14. Jurowski K, Kochan K, Walczak J, Barańska M, Piekoszewski W, Buszewski B (2017) Analytical techniques in lipidomics: state of the art. Crit Rev Anal Chem 47:418–437. https://doi.org/10.1080/10408347.2017.1310613
15. Amaral G, Bushee J, Cordani UG, Kawashita K, Reynolds JH, Almeida FFMDE, de Almeida FFM, Hasui Y, de Brito Neves BB, Fuck RA (2013) Summary for policymakers. In: Climate change 2013—the physical science basis; intergovernmental panel on climate change, vol 369. Cambridge University Press, Cambridge, UK, pp 1–30
16. Baker MJ, Byrne HJ, Chalmers J, Gardner P, Goodacre R, Henderson A, Kazarian SG, Martin FL, Moger J, Stone N (2018) Clinical applications of infrared and Raman spectroscopy: state of play and future challenges. Analyst 143:1735–1757. https://doi.org/10.1039/C7AN01871A
17. Smith E, Dent G (2019) Modern Raman spectroscopy—a practical approach, 2nd edn. Hoboken, NJ, USA, Wiley. ISBN 9781119440581
18. Bergholt MS, Serio A, Albro MB (2019) Raman spectroscopy: guiding light for the extracellular matrix. Front Bioeng Biotechnol 7:303. https://doi.org/10.3389/fbioe.2019.00303
19. Lim NS, Hamed Z, Yeow CH, Chan C, Huang Z (2011) Early detection of biomolecular changes in disrupted porcine cartilage using polarized Raman spectroscopy. J Biomed Opt 16:017003. https://doi.org/10.1117/1.3528006
20. You AYF, Bergholt MS, St-Pierre JP, Kit-Anan W, Pence IJ, Chester AH, Yacoub MH, Bertazzo S, Stevens MM (2017) Raman spectroscopy imaging reveals interplay between atherosclerosis and medial calcification in the human aorta. Sci Adv 3:e1701156. https://doi.org/10.1126/sciadv.1701156
21. Huang N, Short M, Zhao J, Wang H, Lui H, Korbelik M, Zeng H (2011) Full range characterization of the Raman spectra of organs in a murine model. Opt Express 19:22892–22909. https://doi.org/10.1364/OE.19.022892
22. Awan SA, Kibble BP (2007) A universal geometry for calculable frequency-response coefficient of LCR standards and new 10-MHz resistance and 1.6-MHz quadrature-bridge systems. IEEE Trans Instrum Meas 56(2):221–225. https://doi.org/10.1109/TIM.2007.891143
23. Awan SA, Kibble BP, Robinson IA, Giblin SP (2001) A new four terminal-pair bridge for traceable impedance measurements at frequencies up to 1 MHz. IEEE Trans Instrum Meas 50(2):282–285. https://doi.org/10.1109/19.918122
24. Youngs IJ (2000) Dielectric measurements and analysis for the design of conductor/insulator artificial dielectrics. IEE Proc Sci Meas Technol 147:202–208. https://doi.org/10.1049/ip-smt:20000462

25. Dolabella S, Borzì A, Dommann A, Neels A (2022) Lattice strain and defects analysis in nanostructured semiconductor materials and devices by high-resolution X-ray diffraction: theoretical and practical aspects. Small Methods 6(2):2100932. https://doi.org/10.1002/smtd.202100932
26. Ng B, Peng X, Faegh E, Mustain WE (2020) Using nanoconfinement to inhibit the degradation pathways of conversion-metal oxide anodes for highly stable fast-charging Li-ion batteries. J Mater Chem A 8(5):2712–2727. https://doi.org/10.1039/C9TA11708C
27. Chang Q, Ng YXA, Yang D, Chen J, Liang T, Chen S, Zhang X et al (2023) Quantifying the morphology evolution of lithium battery materials using operando electron microscopy. ACS Mater Lett 5(6):1506–1526. https://doi.org/10.1021/acsmaterialslett.3c00065
28. Bashir S, Li R, Song S, Zheng F, Ramirez GA, Houf W, Liu JL (2022) Interactive nanomaterials for energy storage and conversion. In: Nanostructured materials for sustainable energy: design, evaluation, and applications. American Chemical Society, pp 27–81. https://doi.org/10.1021/bk-2022-1421.ch002
29. Babatunde DE, Denwigwe IH, Babatunde OM, Agboola O, Akinsipe GD (2019) Relevance of chemically functionalized nano-fillers and modified nanocomposite in energy systems. In: Polymer nanocomposites for advanced engineering and military applications, 10–65.https://doi.org/10.4018/978-1-5225-7838-3.ch002
30. Wang M, Feng Z (2021) Interfacial processes in electrochemical energy systems. Chem Commun 57(81):10453–10468. https://doi.org/10.1039/D1CC01703A
31. Zheng X, Ahmad T, Chen W (2021) Challenges and strategies on Zn electrodeposition for stable Zn-ion batteries. Energy Storage Mater 39:365–394. https://doi.org/10.1016/j.ensm.2021.04.027
32. Nava-Avendano J, Veilleux J (2017) Plasma processes in the preparation of lithium-ion battery electrodes and separators. J Phys D Appl Phys 50(16):163001. https://doi.org/10.1088/1361-6463/aa6245
33. Wu F, Maier J, Yu Y (2020) Guidelines and trends for next-generation rechargeable lithium and lithium-ion batteries. Chem Soc Rev 49(5):1569–1614. https://doi.org/10.1039/C7CS00863E
34. Chen T, Chen S, Chen Y, Zhao M, Losic D, Zhang S (2021) Metal-organic frameworks containing solid-state electrolytes for lithium metal batteries and beyond. Mater Chem Front 5(4):1771–1794. https://doi.org/10.1039/D0QM00856G
35. Lee JI, Song G, Cho S, Han DY, Park S (2020) Lithium metal interface modification for high-energy batteries: approaches and characterization. Batteries Supercaps 3(9):828–859. https://doi.org/10.1002/batt.202000016
36. Chen S, Xing W, Duan J, Hu X, Qiao SZ (2013) Nanostructured morphology control for efficient supercapacitor electrodes. J Mater Chem A 1(9):2941–2954. https://doi.org/10.1039/C2TA00627H
37. Akhtar S, Lee W, Kim M, Park MS, Yoon WS (2021) Conduction mechanism of charge carriers in electrodes and design factors for the improvement of charge conduction in Li-ion batteries. J Electrochem Sci Technol 12(1):1–20. https://doi.org/10.33961/jecst.2020.01564
38. Abarro JME, Gavan JNL, Loresca DED, Ortega MAA, Esparcia EA Jr, Paraggua JAD (2023) A tale of nickel–iron batteries: its resurgence in the age of modern batteries. Batteries 9(7):383. https://doi.org/10.3390/batteries9070383
39. Tao S, Li M, Lyu M, Ran L, Wepf R, Gentle I, Knibbe R (2022) In operando closed-cell transmission electron microscopy for rechargeable battery characterization: scientific breakthroughs and practical limitations. Nano Energy 96:107083. https://doi.org/10.1016/j.nanoen.2022.107083
40. He SJ, Zhang KQ, Zou YJ, Tian ZH (2022) Three-dimensional printed carbon-based microbatteries: progress on technologies, materials and applications. New Carbon Mater 37(5):898–917. https://doi.org/10.1016/S1872-5805(22)60634-6
41. Wang CM, Li X, Wang Z, Xu W, Liu J, Gao F, Baer DR (2012) In situ TEM investigation of congruent phase transition and structural evolution of nanostructured silicon/carbon anode for lithium ion batteries. Nano Lett 12(3):1624–1632. https://doi.org/10.1021/nl204559u

42. Funke C, Chakrvadhanula VSK (2019) Characterisation of battery materials by electron and ion microscopy techniques: a review. Phys Sci Rev 4(3):20170153. https://doi.org/10.1515/psr-2017-0153
43. Zhang C, Jiang Q, Liu A, Wu K, Yang Y, Lu J, Wang H (2020) The bead-like $Li_3V_2(PO_4)3$/NC nanofibers based on the nanocellulose from waste reed for long-life Li-ion batteries. Carbohyd Polym 237:116134. https://doi.org/10.1016/j.carbpol.2020.116134
44. Wu Y, Ouyang T, Xiong T, Jiang Z, Hu Y, Deng J, Balogun MS (2022) Boosted storage kinetics in thick hierarchical micro–nano carbon architectures for high areal capacity li-ion batteries. Energy Environ Mater 5(4):1251–1259. https://doi.org/10.1002/eem2.12241
45. Han Y, Huang G, Xu S (2020) Structural reorganization-based nanomaterials as anodes for lithium-ion batteries: design, preparation, and performance. Small 16(15):1902841. https://doi.org/10.1002/smll.201902841
46. Orlando A, Franceschini F, Muscas C, Pidkova S, Bartoli M, Rovere M, Tagliaferro A (2021) A comprehensive review on Raman spectroscopy applications. Chemosensors 9(9):262. https://doi.org/10.3390/chemosensors9090262
47. Schmitt R, Nenning A, Kraynis O, Korobko R, Frenkel AI, Lubomirsky I, Rupp JL (2020) A review of defect structure and chemistry in ceria and its solid solutions. Chem Soc Rev 49(2):554–592. https://doi.org/10.1039/C9CS00588A
48. Fleischmann S, Mitchell JB, Wang R, Zhan C, Jiang DE, Presser V, Augustyn V (2020) Pseudocapacitance: from fundamental understanding to high power energy storage materials. Chem Rev 120(14):6738–6782. https://doi.org/10.1021/acs.chemrev.0c00170
49. Pishgar S, Gulati S, Strain JM, Liang Y, Mulvehill MC, Spurgeon JM (2021) In situ analytical techniques for the investigation of material stability and interface dynamics in electrocatalytic and photoelectrochemical applications. Small Methods 5(7):2100322. https://doi.org/10.1002/smtd.202100322
50. Zhao L, Li Y, Yu M, Peng Y, Ran F (2023) Electrolyte-wettability issues and challenges of electrode materials in electrochemical energy storage, energy conversion, and beyond. Adv Sci 10(17):2300283. https://doi.org/10.1002/advs.202300283
51. Márquez-Montesino F, Portales-Martínez B, Martínez-dlCruz L, Vega-Moreno J, Velasco-Arias D, Lemus-Santana AA (2023) Electrodes with excellent capacitance: influence of micro-mesoporosity in derived bituminous carbon. J Energy Storage 73:108934. https://doi.org/10.1016/j.est.2023.108934
52. Saleh HM, Hassan AI (2023) Synthesis and characterization of nanomaterials for application in cost-effective electrochemical devices. Sustainability 15(14):10891. https://doi.org/10.3390/su151410891
53. Qu D, Ji W, Qu H (2022) Probing process kinetics in batteries with electrochemical impedance spectroscopy. Commun Mater 3(1):61. https://doi.org/10.1038/s43246-022-00284-w
54. Liu Y, Jiang SP, Shao Z (2020) Intercalation pseudocapacitance in electrochemical energy storage: recent advances in fundamental understanding and materials development. Mater Today Adv 7:100072. https://doi.org/10.1016/j.mtadv.2020.100072
55. Hu W, Peng Y, Wei Y, Yang Y (2023) Application of electrochemical impedance spectroscopy to degradation and aging research of lithium-ion batteries. J Phys Chem C 127(9):4465–4495. https://doi.org/10.1021/acs.jpcc.3c00033
56. Kim T, Choi W, Shin HC, Choi JY, Kim JM, Park MS, Yoon WS (2020) Applications of voltammetry in lithium ion battery research. J Electrochem Sci Technol 11(1):14–25. https://doi.org/10.33961/jecst.2019.00619

Chapter 5
Synthesis Techniques for Novel Nanomaterials

Abbay Gebretsadik Debalke, Buzuayehu Abebe, and H. C. Ananda Murthy

Abstract Novel nanomaterials (NMs) with high efficiency, low cost, and low energy demand have emerged as an attractive class of materials in high demand for a variety of energy storage applications. They are categorized based on their composition, dimensions, or morphology. For the synthesis of novel NMs, three main approaches are used. Those are physical, chemical, and biological synthesis techniques. In this chapter, various physical synthesis methods such as atomic layer deposition (ALD), chemical vapor deposition (CVD), spray pyrolysis, mechanical milling, electrospinning, sputtering, laser ablation, and microwave-assisted techniques have been explained. In addition, chemical techniques such as solid state, sol–gel, coprecipitation, hydrothermal, microemulsion, molten salt, chemical bath deposition (CBD), and template-assisted synthesis methods of novel NMs have been described in detail. Similarly, green synthesis has also been mentioned as cost-effective, safe and/or non-toxic, easy operation, accessible, and operating under ambient conditions, and we try to compare and contrasts the biological techniques with others.

Keywords Energy storage · Novel nanomaterials · Synthesis techniques

5.1 Introduction

The remarkable class of materials known as nanomaterials (NMs) has drawn a lot of attention because of its unusual characteristics and the fact that it can refer to a wide range of samples with at least one dimension ranging from 1 to 100 nm. Very large surface areas can be achieved through the intentional design of nanoparticles (NPs). NMs have become a major research focus in the realm of affordable

A. G. Debalke (✉) · B. Abebe
Department of Applied Chemistry, Adama Science and Technology University, 1888 Adama, Ethiopia
e-mail: abbaytsadik@gmail.com

H. C. Ananda Murthy
School of Applied Sciences, Papua New Guinea University of Technology, Lae, Morobe Province 411, Papua New Guinea

R. Naik et al. (eds.), *Advanced Nanomaterials for Energy Storage Devices*, Nanostructure Science and Technology, https://doi.org/10.1007/978-3-031-74730-4_5

electrochemical devices for energy conversion and storage applications. Due to their remarkable chemical and physical characteristics, a great deal of research has been done to enhance the efficiency and affordability of electrochemical devices, such as fuel cells, supercapacitors, and batteries. The need for increased productivity, connection, and sustainability at a lower cost is what drives the ongoing development and improvement of these high-performance nanomaterials [1].

Synthesis of novel NMs is the most important step in the research of solid-state chemistry and materials science. There are three classes of novel nanomaterial's synthesis techniques. Those are physical, chemical, and biological methods. Physical methods are synthesis techniques that use the top-down methods, also known as a destructive method, decompose bulk materials into smaller materials, which then transform into NMs [2]. A number of researchers have reported lithography [3], mechanical milling or ball milling [4], laser ablation [5], and sputtering [6] as some examples of the top-down methods. Chemical and biological techniques are considered to be the bottom-up method, also known as the constructive method, involves the building of material from atoms to clusters to nanoparticles [7]. Sol–gel [8], coprecipitation [9], hydrothermal [10], chemical vapor deposition (CVD) [11], spin coating [12], pyrolysis [7], microemulsion [13], chemical bath deposition (CVD) [14], molten salt [15], template-based [16], electro-spinning [17], atomic layer deposition (ALD) [18], and biological synthesis method [19, 20] are examples of bottom-up methods as reported by various researchers. In general, this chapter provides an insight on the synthesis techniques of novel NMs in the applications of energy storage devices since each preparation technique has its own pros and cons.

5.2 Synthesis Techniques

The most crucial stage in materials science and solid-state chemistry research is synthesis. It is possible to prepare novel NMs like glasses, thin films, polycrystalline powders, single crystals, and amorphous samples. Novel NMs can be synthesized using a variety of techniques. Numerous controllable and/or uncontrollable parameters affect each procedure. The parameters impact the size, shape, and phase of individual crystals. The size of the single crystal and its crystallinity are the two most crucial factors in establishing the crystal structure [21]. We will provide an insight for the synthesis techniques of novel NMs in the following sections.

5.2.1 Physical Synthesis Techniques

5.2.1.1 Mechanical Milling

The mechanical milling process is regarded as an economical technique that converts large materials into nanosized ones. This technique can be used to efficiently generate a mixture of various nanophases as well as nanocomposites. For example, Sharma, et al. [22] studied various features of Al–Cu–Mg–T351/MWCNT composites by employing both the mechanical and thermal manufacturing technique. The ball milling procedure has revealed the homogeneous dispersion of CNTs with significantly improved fatigue resistance, creep resistance, ductility, and other mechanical properties of the aluminum-based composites, according to the results. They also provide an amazing synergistic strength. When compared to aluminum-MWCNT composites that were thermally loaded, the mechanical loading of the composite showed improved characteristics [22].

An interesting strategy to improve the energy storage capabilities of lithium iron phosphate, which is utilized in cathodes for lithium-ion batteries, is to reduce the particle size using ball milling. It is still unclear, therefore, how particular storage capabilities and electronic conduction are affected by ball milling settings. The impacts on microstructural, electrical, and electrochemical properties of both solvent-based and dry ball milling (using acetone, methanol, and water) have been examined. Particle agglomeration caused by the use of methanol and water increased the number of pathways for electronic conduction but negatively impacted the specific energy storage capacity. On the other hand, using acetone improved electrical conduction by reducing the creation of these agglomerates. But this also resulted in flaws that prevented an improvement in energy storage. Ultimately, dry milling produced a better specific energy storage capacity, an increase in electronic conductivity, and a more noticeable reduction in particle size [23].

5.2.1.2 Sputtering Technique

Sputter deposition is a thin film deposition technique that uses physical vapor deposition (PVD). The process of sputtering involves blasting solid materials with high-energy particles that are provided by plasma or gas in order to create nanosized materials. It is thought to be a practical method for producing thin films at the nanoscale. Sputtering is a method that physically releases small atom clusters dependent on the energy of the incident gaseous ion by bombarding the target surface with powerful gaseous ions, such as argon ions, as seen in Fig. 5.1. The general sputtering method offers the following benefits: strong adhesion, vast coating area, easy control, low temperature, low damage, high speed, simple equipment, high purity, and low cost. It can be used to create a wide range of nanomaterials, including metals, semiconductors, insulators, etc. [24, 25].

5.2.1.3 Laser Ablation

The method of laser ablation synthesis comprises the production of nanoparticles by shattering the target material with a strong laser beam. With this technique, the precursor or source material was vaporized by high-energy laser irradiation, producing nanosized materials with great homogeneity, purity, and extended

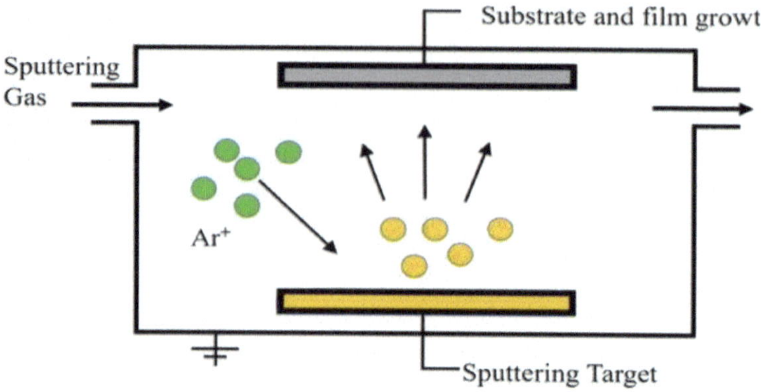

Fig. 5.1 Schematic diagram of magnetron sputtering used for the synthesis of nanoparticles. Reproduced with the permission from Ref. [25] Copyright 2018, IntechOpen

reproducibility. Laser ablation is considered as a green method for producing novel nanomaterials because it does not require any chemical (solvents) or stabilizing agents. Furthermore, in contrast to traditional lengthy approaches, this novel approach provides rapid heating and cooling operations, which enables the production of homogenous nanosystems in a relatively short period of time. By varying processing conditions and sintering mediums, the concentrated laser beam produces fast photothermal and photochemical effects in either air or an inert gas atmosphere, enabling the rapid production of nanomaterials with precise control over geometry, chemistry, crystallinity, and defect density [26].

5.2.1.4 Thermal Decomposition Method

Thermal decomposition method is a synthesis technique of nanomaterials employing heat treatment through endothermic process in which the chemical bonds are broken and divided into smaller ones by heat. The metal is broken down at specific temperatures to form the nanoparticles, which are then formed by a chemical reaction [7].

5.2.1.5 The Arc Discharge Method

This technique can be used to create a variety of nanostructured materials. Fullerenes, carbon nanohorns (CNHs), carbon nanotubes, few-layer graphene (FLG), and amorphous spherical carbon nanoparticles are some of the carbon-based materials produced. This method is extremely important in the production of fullerene nanomaterials. This technique seems to be relatively cheaper and easy to operate. The synthesis requires some process conditions such as; high temperature, the influx of vaporized precursors, buffer gas, and in some cases, a catalyst could be used. These process conditions are allowed to interact in a confined volume of an arc discharge chamber, resulting in the growth of nanomaterial [27, 28].

5.2.1.6 Lithography

It is the process of printing a required shape or structure on a light-sensitive material and selectively removing a portion of the material to create the desired shape and structure. Lithography is a practical method for creating nanoarchitectures with a concentrated electron or light beam. When light is applied onto a photosensitive surface to generate various nanostructured materials, it is termed as photolithography, and when electrons scan a surface covered in sensitive materials to produce nanostructures in the desired locations, it is known as electron beam lithography. The main advantage of nanolithography is its ability to produce a cluster with the desired shape and size from a single nanoparticle. The disadvantages are the requirement for complex equipment and the associated costs [29].

5.2.2 Chemical Synthesis Techniques

5.2.2.1 Sol–Gel Method

It is a wet-chemical synthesis technique that is mostly employed in the fields of ceramic engineering and materials science. It is the process of using water and alcohol-induced polymerization processes to transform a precursor solution into an inorganic solid. First, the precursor must be hydrolyzed in water or dissolved in alcohol in order to form a sol through heating. The next step is condensation, which increases the solvent's viscosity and produces porous structures that are subsequently given time to develop. The development of hydroxo-(M–OH–M) or oxo-(M–O–M) bridges during the condensation or polycondensation process results in the formation of metal-hydroxo- or metal-oxo-polymer formation in solution. In general, condensation produces a gel, while hydrolysis creates a sol, which is essentially a dispersion of colloidal particles in a liquid. This process (the sol–gel method), which is more widely used and has more industrial uses than other currently employed routes, can manufacture large amounts of nanomaterial, especially metal oxide nanoparticles. Morphologies of the developed materials can be depending significantly on the drying method which is selected according to the application in which it will be used [30].

Sol–gel processing technique has been considered as a highly promising method for the synthesis of inorganic and organic–inorganic hybrid nanomaterials due to its impressive features such as the utilization of low processing temperatures ($<$ 100 °C) during synthesis, molecular level composition homogeneity, high purity, better possibility to design the material structure, and the method's ease of control over particle size and shape [31].

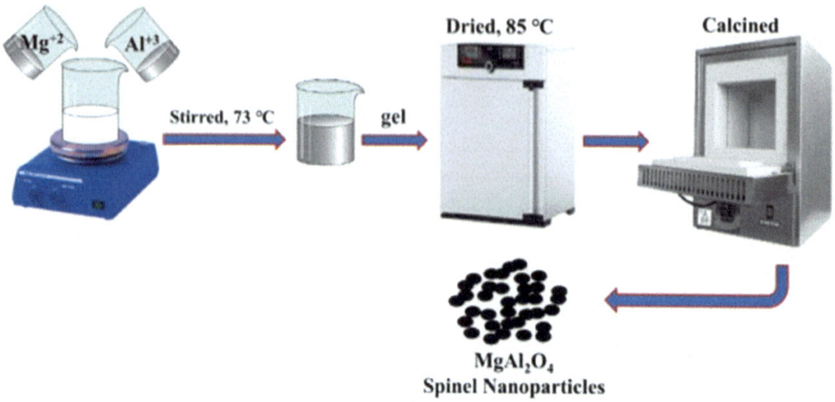

Fig. 5.2 Scheme of $MgAl_2O_4$ spinel NPs preparation through sol–gel process. Reused with the permission from Ref. [32] Copyright 2023, Springer Nature

Figure 5.2 illustrates the effective synthesis of $MgAl_2O_4$ spinel NPs using the sol–gel technique. The acquired data showed that a mesoporous structure was connected to an average crystallite size of about 12 nm and a specific surface area of 108.1 $m^2 g^{-1}$. The obtained $MgAl_2O_4$ spinel NPs were found to be pure by the EDS and FESEM examinations; they were found to be free of contaminants and to have a consistently tiny, uniform, and spherical morphology [32].

This preparation method has also been used to synthesize various porous nanomaterials. For example, TiO_2 nanoparticles were created by Chenari, et al. [33] utilizing the sol–gel process using titanium alkoxide and isopropanol as a precursor. The anatase phase, which has a preferred orientation along the [101] direction, makes up the majority of the particles, as confirmed by the X-ray powder diffraction pattern. The image obtained using scanning electron microscopy clearly displays TiO_2 nanoparticles, which range in size from 60 to 80 nm. This particle cluster, which is made up of aggregated nanoparticles, has a non-ordered, porous structure [33].

5.2.2.2 Coprecipitation Method

Precipitation is the liquid-phase preparation technique that has the benefits such as simple procedure, moderate cost, and well performance of materials. There are three classes of precipitation routes. Those are direct precipitation, coprecipitation, and homogeneous precipitation. Coprecipitation is one of the precipitation techniques that consists of two or more cations in the solution with the same phase. The homogenous precipitation of different components can be obtained by the precipitation reaction. It is a very essential technique to synthesize composite oxide ultrafine samples containing two or more metal components [34].

Spinel NMs and their composites have been synthesized using coprecipitation processes in large quantities. A room temperature chemical coprecipitation approach was successfully used to create cobalt-manganese-nickel oxalates micropolyhedrons. It is interesting to note that flexible solid-state asymmetric supercapacitors have effectively used graphene nanosheets and $Co_{0.5}Mn_{0.4}Ni_{0.1}C_2O_4*nH_2O$ micropolyhedrons as their positive and negative electrode materials, respectively—a capacitive electrode and a battery-type Faradaic electrode. Significantly, the gadget in its assembled state demonstrated a maximum energy density of 0.46 $mWhcm^{-3}$ and operated effectively in both normal and bending situations (0°–180°). In the realm of high-energy–density energy storage devices, the created device configuration is a promising candidate for high-performance flexible solid-state asymmetric supercapacitors because of the resulting device's good flexibility and minimal performance decay even after 6000 cycles [35].

Asma, et al. [36] describe a simple coprecipitation method for synthesizing MXenes (a new class of 2D NMs, which are carbides, nitrides, carbonitrides of transition metals) with copper chromite (($Ti_3C_2T_x$)/$CuCr_2O_4$) nanocomposites, which can be used to study special and important energy storage properties. This will enable researchers to design and create nano-electrode materials with the highest possible specific capacitance, conductivity, and stability. The average size of the crystallites was determined to be 8.7 nm, whereas the SEM micrographs clearly show a sandwich-like morphology with an average grain size of 1.53 nm.

In other work, Gupta et al. [37] synthesized nanocrystalline titanium dioxide (TiO_2) NPs utilizing three distinct methods, and they used these nanoparticles for various applications. Through the application of chemical techniques, the calculated size of the nanoparticle was less than 11 nm. All of the samples' XRD patterns had prominent, broad peaks, indicating that they were all in the nanocrytalline range. The synthesized titania particles were pure anatase, with sizes ranging from 7 to 11 μm, after employing the sol–gel method with Titanium tetra isopropoxide (TTIP) as a precursor. The size of titania microparticles can be reduced up to 60 nm through the use of solid-state mechanical reduction, but their phase stays the same. Moreover, it was noted that Titania produced by the coprecipitation process using $TiCl_4$ as a precursor was entirely rutile and ranged in size from 3 to 7 nm. Consequently, these findings verify that the coprecipitation technique has proven to be most effective in producing smaller nanoparticles. It has been observed that using the right synthesis technique, the properties of TiO_2 can be adjusted.

5.2.2.3 Microemulsion Method

A chemical process known as "micro-emulsions" are dispersions of two immiscible liquids that are homogenous in macroscale and microheterogeneous in nanoscale, with nanosized domains of one or both liquids in the other sustained by an interfacial coating of molecules that are surface active. Particle size and stability are the key differences between a standard emulsion and a microemulsion. Emulsions that are normally aged undergo Ostwald ripening and droplet coalescence. Microemulsions often form spontaneously and are solitary optically isotropic, and thermodynamically stable. Microemulsions are characterized by their enormous interfacial area, ultralow interfacial tension, and ability to solubilize both soluble and insoluble substances. Microemulsions contain surfactants and are characterized as either water-in-oil (W/O), oil-in-water (O/W), or intermediate bicontinuous structural kinds that can reversibly change from one type to another based on the percentage of different components and the hydrophilic–lypophilic balance (HLB) value of the utilized surfactant. Monodispersed droplets with a size range of 5–100 nm make up the dispersed phase. By changing the relevant parameters, such as the type of stabilizer, continuous phase, the amount of precursor dissolved in the nanodroplets, and the water content—also known as the molar ratio of water to surfactant—the size of the nanodroplets can be altered. Furthermore, the stability of the microemulsion can be impacted by temperature, pressure, reagent concentration, and salt addition. Microemulsion has shown to be incredibly adaptable and consistent, allowing for precise control over nanoparticle size and producing particles with a narrow size distribution [38, 39].

5.2.2.4 Molten Salt Method

Molten salt synthesis (MSS) is a productive technique for producing a large variety of nanomaterials, especially metal oxides. The MSS method is a nonaqueous process that enables the synthesis of comparatively thermally and chemically stable metal oxides; extremely reactive metal oxide precursors have also been employed because extra solvents like water is not present. Its simplicity, greenness, dependability, scalability, and generalizability provide it advantages over other approaches. Zuniga et al. [40] proposed the MSS technique for the effective manufacture of complex metal oxide nanoparticles (NPs), using pyrochlore lanthanum hafnium oxide ($La_2Hf_2O_7$) as a sample. Additionally, by adjusting several synthesis parameters like pH, temperature, time, and post-annealing, this approach has the unique capacity to create NPs with distinct material properties. The authors are able to create extremely crystalline, non-agglomerated, and homogeneous NPs by adjusting these parameters. Furthermore, the authors can further investigate the impact of particle size on different attributes by adjusting the concentration of the ammonium hydroxide solution utilized in the MSS process, which alters the particle size of the $La_2Hf_2O_7$ NPs [40].

MSS method has also been highlighted as being scalable for the production of 2D oxides within a short reaction time ($\sim$ 1 min). It has the ability to produce NPs with a relatively large surface area, mesoporosity, and good thermal stability and thin, flake-like morphology. The thickness of the synthesized flakes can be controlled by adjusting the salt and precursor molar ratio. Although through the MSS route, highly crystalline materials with controlled facets can be synthesized; the high operating temperature comprises the main drawback considering potential industrial realization. The nature of salt solid crystals has lately been highlighted as being crucial to a unique synthesis strategy, in addition to the impact of free ions in the molten salt route [41].

Lasfargues et al. [42] described the in situ synthesis of titanium dioxide nanoparticles for use in heat transfer fluid and thermal energy storage utilizing $TiOSO_4$ precursor in molten salt phase. The specific heat capacity of a binary eutectic mixture of potassium and sodium nitrate (60/40) was found to be significantly increased by TiO_2 nanoparticles (average particle size 16 nm) produced in molten salt phase in this study. This development of a potentially scalable alternative production method is necessary because the current methods are not economically viable for large quantities. Improving molten salt's specific heat capacity would encourage more competitive pricing for concentrating solar power plant electricity production [42].

5.2.2.5 Hydrothermal Method

Hydrothermal synthesis is a specialized process that combines high temperatures and pressure to make nanomaterials. It works particularly well at producing metal oxide nanoparticles and nanowires. This approach is appealing because it can carefully control the structure and development of crystals, giving materials unique properties. Sharp reactions that are produced by extremely high temperatures promote the creation of nanoscale structures with enhanced energy storage capabilities. This method has the potential to create more powerful and efficient energy storage devices, such as supercapacitors and batteries, and it holds enormous promise for advancing energy storage technologies and clearing the way for a sustainable and technologically advanced future [43].

This method demonstrated distinct benefits in comparison with others. At very high temperatures, it can form materials in the nanoscale that are unstable. Additionally, it can produce high vapor pressure, nanoscale compounds with very little loss of beginning ingredients. The hydrothermal process is useful for creating nanomaterials with various morphologies, including nanospheres, nanowires, nanosheets, and nanorods [44].

A novel multinary nanocomposite $GO/AlCrO_3/Mn_3O_4/SiO_2/SnO_2$ was synthesized through hydrothermal method and served as an electrode material for energy storage. The findings show the emergence of a multilayered, complex structure with a comparatively high aspect ratio, indicating that the quasi-spherical agglomerated morphology [45].

A rutile phase of titanium dioxide was also produced by using the hydrothermal method for 20 h at 180 °C. The produced titanium dioxide was found to have a hierarchical microsphere structure, ordered by nanorods, with an average uniform size of approximately 1–3 μm, as demonstrated by SEM pictures. The nanorods' particle diameter measured 62 nm. The ability to synthesize titanium dioxide with both rutile and anatase phases at medium temperatures and reaction times is another benefit of this method. The constituent contents of Ti 61.03% and O 39.997% were found by good uniform distributed EDS analysis of the TiO_2 sample, which demonstrated good features. This synthesis process offers a straightforward way to create nanostructured titanium dioxide, which has numerous applications [46].

Furthermore, hydrothermal synthesis can only occur under supercritical conditions and is unaffected at temperatures and pressures below the critical threshold for a particular solvent, at which variations between liquid and vapor vanish. Temperature, reaction duration, and alkaline concentration all affect the hydrothermal process. The development of crystals and the nucleation of nanoparticles are both aided by temperature. The yield, length, and degree of crystallinity of nanotubes all rise with temperature, with 100–200 °C being the ideal range. Yield rises with hydrothermal time; nevertheless, extended hydrothermal time causes morphological changes in nanoparticles, such as the conversion of nanotubes to nanofibers [47].

In addition, Bregadiolli et al. [48] examined the effects of various environmental factors on the morphology and structures of TiO_2 that had been synthesized in acidic and alkaline media using microwave-assisted hydrothermal synthesis. The temperatures were low—below 150 °C—and the synthesis timeframes were short—between 2 and 60 min. The findings showed that there are no appreciable changes in the material's structure with reaction time or temperature in the acidic synthesis. Nonetheless, under fundamental circumstances, the particle development and morphology are significantly influenced by the Na^+ ion concentration. In acidic conditions, the morphology of the nanoparticles displays irregular spheres, whereas in an alkaline media, needle-like structures and aggregated nanotube-like structures are created in a matter of minutes. In both systems, high surface area was attained, despite the morphology and structure differing [48].

5.2.2.6 Soft and Hard Templating Methods

Both soft and hard template techniques are heavily used in the manufacture of nanoporous materials. In hard templating techniques, the precursor molecules are pumped into stiff molds to produce the final nanoporous oxide product. The rigid and porous template structure is impregnated with a mixture of metal precursors and chelating agents, as shown in Fig. 5.4 [49]. The templates are then removed by calcining and etching the structure using an acidic or alkaline aqueous solution, yielding the required nanoporous oxide materials. Because hard templating creates ordered and stable nanoporosity structures and the stiff template can withstand high temperatures, it can be used for a wide range of materials with high crystallinity.

The hard template method's disadvantages, however, include a decreased yield due to the inability to fine-tune the pore size. In addition, there are still drawbacks such as the longer time required to fully incorporate metal precursors into the template and the challenge of fully eliminating the hard template with an acid or alkaline solution. The purpose of using soft templates in the creation of nanoporous oxide materials has been extensively documented. The idea is to co-assemble inorganic species and surfactant molecules into ordered structures that are produced after the template is removed [49]. Amphiphilic chemicals, which have hydrophobic and hydrophilic counterparts, lower surface and interfacial tension and are used as surfactants for soft templating. Controlling the morphology, size, and geometric structure is possible with the soft template method, which is not possible with the hard-template methods. However, the main challenges with soft-templating are the limited thermal stability of the soft templates and the restriction of crystal development through heat treatment [50].

5.2.2.7 Atomic Layer Deposition (ALD) Technique

Atomic layer deposition (ALD) is a thin film deposition technology that creates sub-monolayers of film by direct chemical reaction using a series of chemical precursors applied one after the other to a substrate's surface. As its name suggests, it is essentially atomic in nature and produces accurate atomic-scale layer deposition on the surface of a chosen substrate. Furthermore, the uniform deposition of conformal films with controlled thickness at the Angstrom level has been achieved using ALD, a vapor phase deposition technique, even on intricate three-dimensional surfaces. This approach can also result in increased electrical device efficiency. This technology has attracted significant interest both for fundamental understanding of how new functional materials can be synthesized by ALD and for numerous practical applications, including energy storage systems. This method has garnered substantial interest for a variety of real-world applications, such as energy storage devices, as well as for a fundamental understanding of how new functional materials can be created using ALD [51].

However, potential surface contamination may have an impact on the layer's quality and stickiness that the ALD deposits. For this reason, prior to the ALD procedure, each substrate was carefully cleaned in an ultrasonic cleaner and then dried [52]. Figure 5.3 illustrates the deposition process of tin nitride (SnNx) on the surface of the substrates by using ALD.

5.2.2.8 Chemical Vapor Deposition (CVD)

CVD is a versatile technique in which components can be deposited onto a substrate while they are in the vapor phase. It facilitates the precise synthesis of nanowires and thin films from various materials. CVD plays a crucial role in the realm of energy storage by enabling the production of homogenous and meticulously built nanomaterials. These tailored nanostructures allow for rapid charge and discharge rates, which significantly improve the performance of batteries and capacitors. With CVD, material properties and structures may be highly controlled, leading to more advanced and efficient energy storage devices. This encourages industry innovation [54].

Furthermore, the creation of nanoscale carbon-based materials, such as carbon nanotubes, is greatly aided by this synthesis method. In CVD, the chemical reaction of vapor-phase components results in a thin coating on the substrate surfaces. High

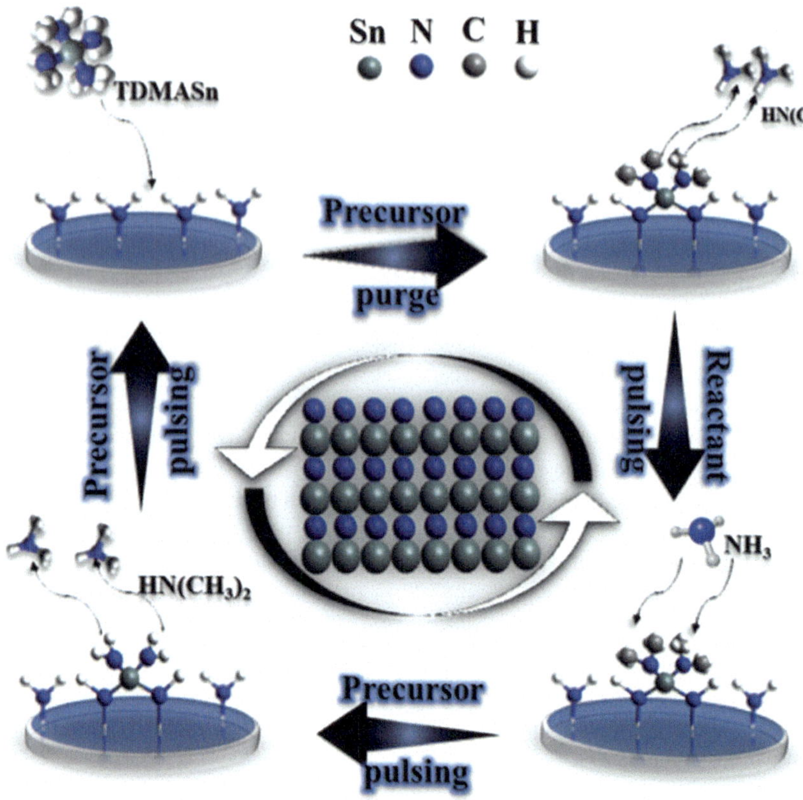

Fig. 5.3 A prototype ALD method employing TDMASn and NH_3 as precursors to produce tin nitride (SnNx). Reused with the permission from Ref. [53] Copyright 2023, John Wiley and Sons

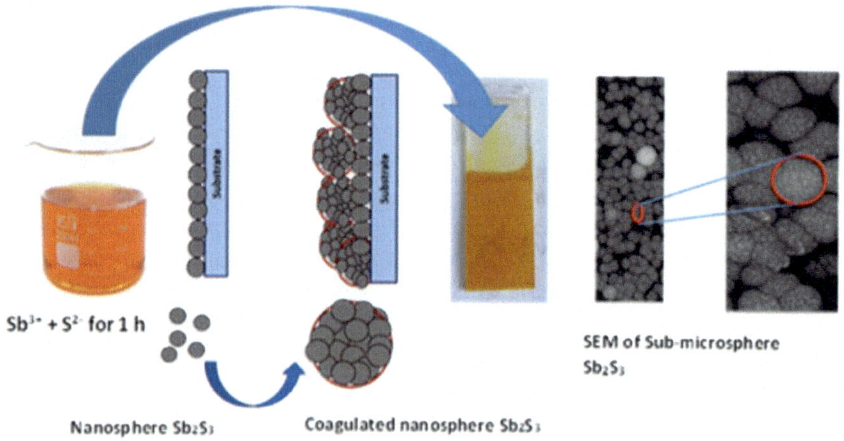

Fig. 5.4 Schematic illustration of the preparation of nanosphere Sb_2S_3 layers. Reused with the permission from Ref. [63] Copyright 2023, RSC Advanced

chemical purity, nonhazardousness, volatility, good stability during evaporation, long shelf life, and low cost are all qualities that the precursor should meet. Additionally, the decomposition process produced no leftover contaminants [55].

5.2.2.9 Electrospinning

It is regarded as a versatile production technique used to produce nanofiber-based polymers such as polyurethane nano-fibers and continuous nanoporous oxide fibers with a variety of compositions. The fundamental electrospinning system consists of a high-voltage system, needles, and a collector [56]. An exposed liquid droplet is electrified and subsequently extended toward the opposing collector when an electric field is generated between two electrodes. The qualities of the electrospun fibers can be adjusted using a variety of process parameters. For instance, the type of solvent or polymer used in the solution, the reaction temperature, the relative humidity, the electrical polarity applied to the nozzle, the distance between the nozzle and the collector, and the applied voltage, and all of these factors have a significant impact on the production of the final product [57].

5.2.2.10 Spray Pyrolysis Method

A deposition method known as spray pyrolysis has been used to produce fine dispersive porous or thick powders or films. In order to create fumed silica, the first flame synthesis (FS) reactors for nanoparticles were developed in the 1940s. The first principles of the FS technique were published in 1971 by G. D. Ulrich. The first substances to be created using flame synthesis and granted patents were silica and titanium dioxide [58].

The spray pyrolysis technique consists of three main stages: the synthesis process, aerosol formation and transport, and the composition of the precursor solution. Each of these steps is adjusted based on the ultimate chemical and physical properties of the material being targeted; these modifications, together with the materials and procedures selected at each step, have an impact on the subsequent steps. As a result, in the first stage, the precursor solution's chemical composition must contain one or more compounds that, upon pyrolysis, will provide the necessary chemical composition. The optimal choice for the aerosol generation/transport procedure, temperature, and synthesis rate will all depend on the solvent used, which will also limit the maximum concentration of the precursor component in the solution. In the subsequent phase, the final material's morphological properties and the appropriate range of synthesis temperatures will be established by the aerosol droplet size distribution, which is ascertained by the aerosol formation process. The kind and flow rate of the carrier gas will either increase or decrease the likelihood of a reactive interaction with the precursor molecule. Whether the material generated is a powder or a film coating will depend on whether the last chemical reaction occurs on a hot substrate or in a gas phase at the end. In general, the concentration molarity of the precursor solution, the carrier gas flow rate, and the synthesis temperature are the synthesis parameters that matter most when considering an experimental setup [59].

For instance, Widiyandari et al. [60] used flame spray pyrolysis (FSP) in a single step to produce titanium dioxide (TiO_2) tiny particles with a narrow size distribution. LPG (liquid petroleum gas) was used as the fuel source, and titanium (IV) isopropoxide was used as the precursor. TiO_2 was found to have a consistent spherical shape and a particle size of between 200 and 500 nm. The size distribution changed toward the larger size as the precursor concentration increased. The TiO_2 product's crystalline structure exhibited a significant specific surface area and displayed a mixed phase of rutile and anatase [60].

Adhesion to the substrate is crucial for the quality of the deposited layer. It can be possible to manage a number of pyrolysis process parameters, including the density, crystal phases, size of the spray droplets, and chemical character of the products that are produced. When producing multicomponent materials, the process of selection and precursor preparation becomes tricky. Due to the fact that sprays are susceptible to fluctuations in the liquid fuel supply, which may have an impact on the growth conditions of nanoparticles, the precursor should be supplied into the reactor slowly. Particle diameter rises with increasing precursor flow rate but falls with increasing dispersion gas flow rate due to reactions and oxidizers mixing quickly [61].

This technique has several important advantages, such as the ability to fabricate entire multilayer devices by deposition of different functional layers in the same chamber, and the ability to produce uniform and dense films with desirable crystallinity by multiple repetitions of spraying and/or annealing cycles. The manufacture of thin films and NPs via spray pyrolysis techniques eliminates the need for lengthy drying times or additional purification steps, which could raise the overall cost and thermal budget of producing the desired NMs. In addition to the benefits already discussed, this method has certain drawbacks. These include the requirement for temperature control and the challenge of obtaining low-temperature allotropic forms of the corresponding products; therefore, cooling systems and precise temperature control are necessary for large-scale production [62].

5.2.2.11 Chemical Bath Synthesis

A chemical bath deposition (CBD) is a technique employed to synthesize nanosphere-structured materials like antimony sulfide layer through which the cleaned substrates are vertically dipped in the solution bath at a low deposition temperature and optimized pH. The nanosphere layers are formed on the substrate surface due to the adsorption and nucleation of the materials on the substrate illustrated in Fig. 5.4 [63].

Moreover, a wet chemical synthesis technique called laser-assisted chemical bath synthesis (LACBS) uses laser irradiation to encourage the controlled precipitation of nanostructures like zinc oxide (ZnO). This process is not the same as laser ablation, which is a strictly physical process that uses a concentrated laser beam to remove material from a solid surface. The advantages of chemical bath synthesis and laser irradiation are combined in LACBS, which improves control over the size, shape, and crystallinity of the resultant nanostructures. The basis of this method is the photothermal effect that a concentrated laser produces, which produces an incredibly controlled localized temperature field [64]. ZnO nanostructures are grown by LACBS with the help of laser irradiation and chemical bath synthesis. A chemical bath is created by dissolving zinc salts and complexing agents, and then exposing the material to laser radiation to accelerate the nucleation process and produce homogenous ZnO nuclei that grow into bigger structures. Factors including concentration, temperature, pH, and reaction time impact the morphology, size, and orientation of the nanostructure in addition to the laser irradiation parameters [65].

5.2.2.12 Microwave-Assisted Method

Microwaves, which are electromagnetic waves with frequency ranging from 0.3 to 300 GHz and wavelengths between 1 mm and 1 m, are used in microwave-assisted techniques. Heat is produced in microwaves by the spinning, friction, and collision of polar molecules as they attempt to align themselves with the fast-varying alternating electric field. If ions are present in the solution, they will flow through it while continuously shifting their paths in response to the electric field's orientation, raising the local temperature as a result of friction and collision. When compared to traditional heating methods, microwave heating offers an alternate heat source for quick heating with a shorter reaction time and a higher reaction rate, excellent selectivity, uniform heating, and higher yield [66, 67].

Heating using microwaves can be done in two different ways: continuously and with pulses. Assuming that the kinetics or mechanism of the chemical reaction remains unchanged despite a significant reduction in reaction time for a microwave-induced reaction, the first mechanism suggests that the thermal heating impact is responsible for the increased reaction rate. As a result, the effects of microwave irradiation in chemical reactions are caused by both thermal and nonthermal effects. The second explanation that has been suggested assumes the existence of "nonthermal microwave effects" in addition to thermal effects; hence, microwave irradiation's effects on chemical reactions are caused by both thermal and nonthermal effects. The direct contact between particular molecules in the reaction media and microwaves is what causes the nonthermal effects [66].

For this synthesis technique to be effective, the nanomaterials produced must possess features that allow for the effective absorption of microwave energy. In order to create a composite consisting of highly conductive, stable $Ti_3C_2T_X$ MXene and nanospheres $Ni_{0.5}Co_{0.5}Se_2$, Widiyandari et al. [68] devised a straightforward and effective microwave-assisted synthesis process utilizing the materials' microwave-absorbing qualities. The results show that in just a short period of time, 3D $Ni_{0.5}Co_{0.5}Se_2$ bimetallic selenide nanospheres were uniformly loaded within the few-layer $Ti_3C_2T_X$ MXene wrapper. In general, the few-layer $Ti_3C_2T_X$

MXene structure allows for the direct growth of selenide thanks to the microwave-assisted synthesis process, which may offer fresh insights into the production of high-pseudocapacitance supercapacitors [68].

Synthesis parameters, such as precursor identity, reaction time and temperature, precursor's molar ratios, and solvent mixtures, have a great impact on the nature of the developed NMs. For example, using thioacetamide (TAA) as the sulfur precursor, a series of FeS_2 (pyrite) samples were first created to produce a unique, consistent octahedra-based morphology. When the sulfur precursor was changed from TAA to L-cysteine, the resulting changes in morphology from octahedral to nanosheet aggregates and chemical composition from FeS_2 to an iron thiospinel structure, Fe_3S_4 (greigite), were observed [69].

5.2.3 Green Synthesis Techniques

Green synthesis provides an excellent substitute for conventional techniques for the production of different nanostructured materials. Many bio-reductants obtained from microorganisms, plants, and agricultural wastes have been used to manufacture a variety of metals and metal oxide nanoparticles [70, 71]. Due to the ability of microorganisms to reduce metal ions and create NPs, microbial-mediated synthesis has demonstrated considerable potential. Additionally, since fungi and algae contain bioactive ingredients like phycobilin, flavonoids, glucosides, tannins, and saponins that are used as capping and reducing agents, phenolics, and chlorophyll, they have emerged as viable alternatives that offer scalable and sustainable methods for producing nanoparticles [72, 73]. The variety and abundance of plant species offer a rich source of bioactive molecules for the synthesis of NPs, which has drawn a lot of attention to plant-mediated synthesis. The capacity of plant components like leaves, roots, and seeds to reduce metal ions and promote the creation of NPs has been investigated [74, 75]. Furthermore, there have been new reports on the green synthesis of NPs from biomass and waste by using fungus, plants, microbes, and agro-industrial bio-wastes [76, 77].

In general, when compared to traditional synthesis, green synthesis offers up to 40% cost savings, a 30% reduction in energy usage, and a 50% boost in industrial output. Additionally, the green synthesis of nanoparticles has drawn more interest as a low-cost, non-toxic, and ecologically acceptable technique [78]. Furthermore, this synthesis method operates under ambient conditions and is not only safe and benign but also affordable, easily accessible, and uncomplicated. These biological agents perform the double functions of capping and reducing agents by stabilizing the nanoparticles and accelerating the reduction process [79].

5.3 Summary

Nanomaterials have gained significant attention as a remarkable class of materials due to their unique properties and the fact that they encompass a wide range of samples with at least one dimension ranging from 1 to 100 nm. The deliberate design of novel nanomaterials enables the achievement of extremely large surface areas. In the field of cost-effective electrochemical devices for energy storage and conversion applications, nanomaterials have emerged as a key area of research. Their exceptional physical and chemical properties have led to extensive investigations aimed at improving the performance and cost-effectiveness of electrochemical devices, including batteries, supercapacitors, and fuel cells. The continuous development and enhancement of these high-performance materials are driven by the demand for enhanced productivity, connectivity, and sustainability at a reduced cost.

Synthesis of these novel nanostructured materials is the most important part of solid-state chemistry research and in materials science. There are different synthesis methods for various novel nanomaterials which are controlled by several controllable and/or uncontrollable parameters. The parameters influence the obtaining of phase, morphology, and size of single crystals.

There are a lot of physical synthesis techniques for novel NMs that have been used for energy storage devices. The mechanical milling synthesis method is regarded as one of an economical technique that converts bulk materials into nanosized ones. Sputtering techniques are thin-film deposition techniques that use PVD which involves blasting solid materials with high-energy particles that are provided by plasma or gas in order to create nanosized materials. The method of laser ablation synthesis comprises the production of NPs by shattering the target material with a strong laser beam for producing nanosized materials with great homogeneity, purity, and extended reproducibility. Thermal decomposition method is a synthesis technique of NMs employing heat treatment through endothermic process in which the chemical bonds are broken and divided into smaller ones by heat. The arc discharge technique can be highly used to create a variety of carbon-based nanostructured materials such as fullerenes, carbon nanohorns, carbon nanotubes, few-layer graphene, and amorphous spherical carbon NPs. Lithography is the process of printing a required shape or structure on a light-sensitive material using a concentrated electron or light beam and selectively removing a portion of the material to create the NMs with the desired shape and structure.

The following are the most common chemical synthesis methods of various NMs. Sol–gel is a wet-chemical synthesis technique that involves water and alcohol-induced polymerization processes to transform a precursor solution into solid NMs. Coprecipitation is a very essential technique to synthesize composite oxide ultra-fine samples containing two or more metal components through precipitations. A chemical process known as "micro-emulsions" are dispersions of two immiscible liquids that are homogenous on macroscale and microheterogeneous in nanoscale, with nanosized domains of one or both liquids in the other sustained by an interfacial coating of molecules that are surface active. MSS method is a nonaqueous process that enables the synthesis of thermally and chemically stable metal oxides through the use of extremely reactive metal oxide precursors since extra solvents like water are not present. Hydrothermal synthesis is a specialized process that combines high temperatures and pressure to make NMs. Soft and hard template techniques are heavily used in the manufacture of nanoporous materials. ALD is a thin film deposition technique that creates sub-monolayers of film by direct chemical reaction using a series of chemical precursors applied one after the other to the surface of a substrate. CVD is a versatile preparation technique in which components can be deposited onto a substrate while they are in the vapor phase to form nanowires and thin films precisely. Spray pyrolysis is a deposition method that has been used to produce fine dispersive porous or thick powders or films. CBD is a technique employed to synthesize nanosphere-structured materials through which the substrates are vertically dipped in the solution bath at specified conditions. Microwave-assisted methods are a synthesis method that uses microwave radiation as an alternative heat sources with frequencies which range from 0.3 to 300 GHz and with wavelengths between 1 mm and 1 m.

Finally, green synthesis provides an excellent substitute for conventional techniques in the synthesis of diverse nanostructured materials through the utilization of biological components obtained from plant extracts, waste materials, and microorganisms. This method works under ambient conditions and is benign, safe, economical, scalable, and simple. These biological agents play the dual roles of capping and reducing agents by stabilizing the nanoparticles and accelerating the reduction process.

References

1. Saleh HM (2023) Synthesis and characterization of nanomaterials for application in cost-effective electrochemical devices. Sustainability 15(10891):1–52
2. Ijaz I, Gilani E, Nazir A, Bukhari A (2020) Detail review on chemical, physical and green synthesis, classification, characterizations and applications of nanoparticles. Green Chem Lett Rev 13(3):59–81. https://doi.org/10.1080/17518253.2020.1802517
3. Martín-Jimeno FJ et al (2017) A 'nanopore lithography' strategy for synthesizing hierarchically micro/mesoporous carbons from ZIF-8/graphene oxide hybrids for electrochemical energy storage. ACS Appl Mater Interfaces 9(51):44740–44755. https://doi.org/10.1021/acsami.7b16567
4. Yang H et al (2023) Recent advances in ball-milling-based silicon anodes for lithium-ion batteries. Energies 16(7). https://doi.org/10.3390/en16073099
5. Li L et al (2021) Laser irradiation construction of nanomaterials toward electrochemical energy storage and conversion: ongoing progresses and challenges. InfoMat 3(12):1393–1421. https://doi.org/10.1002/inf2.12218
6. Wang F et al (2020) Magnetron sputtering enabled synthesis of nanostructured materials for electrochemical energy storage. J Mater Chem A 8(39):20260–20285. https://doi.org/10.1039/d0ta07644a
7. Mekuye B (2023) Nanomaterials: an overview of synthesis, classification, characterization, and applications, vol 4, pp 486–501. https://doi.org/10.1002/nano.202300038

8. Zhang XL et al (2023) Sol–gel method preparation and high-rate energy storage of high-entropy ceramic (FeCoCrMnNi)C porous powder. Ceram Int 49(17):29327–29338. https://doi.org/10.1016/J.CERAMINT.2023.06.228
9. Qiu Q, Wang J, Yao P, Li Y (2023) A facile coprecipitation approach for synthesizing $LaNi_{0.5}Co_{0.5}O_3$ as the cathode for a molten-salt lithium–oxygen battery. Faraday Discuss 248:327–340. https://doi.org/10.1039/d3fd00078h
10. Hussain M et al (2024) Hydrothermal synthesis of Nd-doped $FeTiO_3$ perovskite electrode for enhanced energy storage applications. J Energy Storage 84(PA):110920. https://doi.org/10.1016/j.est.2024.110920
11. Fang Z, Zhou P, Tian Y, Fang H, Zhang Q (2024) One-step chemical vapor deposition synthesis of Si NWs@C core/shell anodes without additional catalysts by the oxide-assisted growth mechanism for lithium-ion batteries. Dalt Trans 53(21):9052–9061. https://doi.org/10.1039/D4DT00131A
12. Darvina Y, Desnita D, Zainul R, Laghari IA, Azril A, Abdullah M (2024) Advancements in Fe_3O_4/PANI nanocomposite film technology: synthesis and characterization. J Med Chem Sci 7(5):689–702. https://doi.org/10.26655/JMCHEMSCI.2024.5.5
13. Ali AM (2019) Corresponding author preprint not peer reviewed PRT not peered. Int J Adv Sci Res Dev 1(3394):7–20
14. Geldasa FT, Kebede MA, Werta SZ, Hone FG (2024) Chemical bath deposition approach to produce three different morphologies of PbO thin films from different cation concentrations. Indian J Phys. https://doi.org/10.1007/s12648-024-03083-w
15. Gandla D, Zhuang Z, Jadhav VV, Tan DQ (2023) Lewis acid molten salt method for 2D MXene synthesis and energy storage applications: a review. Energy Storage Mater 63. https://doi.org/10.1016/J.ENSM.2023.102977
16. Gharat S et al (2024) A review on template synthesis of nanoparticle. Int J Pharm Pharm Sci 16(5):22–29. https://doi.org/10.22159/ijpps.2024v16i5.50661
17. Sun G, Sun L, Xie H, Liu J (2016) Electrospinning of nanofibers for energy applications. Nanomaterials 6(7). https://doi.org/10.3390/nano6070129
18. Karimzadeh S, Safaei B, Yuan C, Jen TC (2023) Emerging atomic layer deposition for the development of high-performance lithium-ion batteries, vol 6, no 1. Springer Nature, Singapore. https://doi.org/10.1007/s41918-023-00192-8
19. Abuzeid HM, Julien CM, Zhu L, Hashem AM (2023) Green synthesis of nanoparticles and their energy storage, environmental, and biomedical applications. Crystals 13(11). https://doi.org/10.3390/cryst13111576
20. Powell MD, LaCoste JD, Fetrow CJ, Fei L, Wei S (2021) Bio-derived nanomaterials for energy storage and conversion. Nano Sel 2(9):1682–1706. https://doi.org/10.1002/nano.202100001
21. Ben Smida Y, Marzouki R, Kaya S, Erkan S, Faouzi Zid M, Hichem Hamzaoui A (2020) Synthesis methods in solid-state chemistry. Synth Methods Cryst. https://doi.org/10.5772/intechopen.93337
22. Sharma S, Patyal V, Sudhakara P, Singh J, Petru M, Ilyas RA (2021) Mechanical, morphological, and fracture-deformation behavior of MWCNTs-reinforced (Al–Cu–Mg–T351) alloy cast nanocomposites fabricated by optimized mechanical milling and powder metallurgy techniques. Nanotechnol Rev 11(1):65–85. https://doi.org/10.1515/ntrev-2022-0005
23. Sierra-Uribe JH et al (2024) Impact of ball milling on the energy storage properties of $LiFePO_4$ cathodes for lithium-ion batteries. J Solid State Electrochem: 1–9. https://doi.org/10.1007/S10008-024-05927-1/METRICS
24. Chodun R et al (2022) The sputtering of titanium magnetron target with increased temperature in reactive atmosphere by gas injection magnetron sputtering technique. Appl Surf Sci 574:151597. https://doi.org/10.1016/j.apsusc.2021.151597
25. Shi F (2018) Basic theory of magnetron sputtering 1. Principle of magnetron sputtering, pp 1–5
26. Devina W et al (2023) Rapid synthesis of nanomaterials by solvent-free laser irradiation for energy storage and conversion. Chem Phys Rev 4(4). https://doi.org/10.1063/5.0150598

27. Awoji MO, Onoja AD, Echi MI (2023) Synthesis of graphene via arc discharge and its characterization: a comparative approach. East Eur J Phys 2023(1):252–257. https://doi.org/10.26565/2312-4334-2023-1-34
28. Tepale-Cortés A, Moreno-Saavedra H, Hernandez-Tenorio C, Rojas-Ramírez T, Illescas J (2021) Multi-walled carbon nanotubes synthesis by arc discharge method in a glass chamber. J Mex Chem Soc 65(4):480–490. https://doi.org/10.29356/JMCS.V65I4.1486
29. Stokes K, Clark K, Odetade D, Hardy M, Goldberg Oppenheimer P (2023) Advances in lithographic techniques for precision nanostructure fabrication in biomedical applications, vol 18, no 1. Springer, US. https://doi.org/10.1186/s11671-023-03938-x
30. Bokov D et al (2021) Nanomaterial by sol-gel method: synthesis and application. Adv Mater Sci Eng 2021:1–21
31. Catauro M, Ciprioti SV (2021) Characterization of hybrid materials prepared by sol-gel method for biomedical implementations. A critical review. Materials (Basel) 14(7). https://doi.org/10.3390/ma14071788
32. Eslami A, Abdollah S, Maryam L, Maryam S, Morteza K (2024) Sol-gel synthesis, characterization, and electrochemical evaluation of magnesium aluminate spinel nanoparticles for high-capacity hydrogen storage, pp 215–225. https://doi.org/10.1007/s10971-023-06260-1
33. Chenari HM, Seibel C, Hauschild D, Reinert F, Ave N, Hubland A (2016) Titanium dioxide nanoparticles: synthesis, X-ray line analysis and chemical composition study 2. Exp Methods 19(6):1319–1323
34. Coprecipitation O, Xiao Y, Sun T, Zhao Y (2020) Experimental study on preparation of ferropericlase. Minerals 2(2018):179
35. Zhang Y et al (2015) Room temperature synthesis of cobalt-manganese-nickel oxalates micropolyhedrons for high-performance flexible electrochemical energy storage device, vol 5, no 8536, pp 1–6. https://doi.org/10.1038/srep08536
36. Asma A et al (2023) Fabrication and characterization of MXene/$CuCr_2O_4$ nanocomposite for diverse energy applications fabrication and characterization of MXene/$CuCr_2O_4$ nanocomposite for diverse energy applications. J Mater Res Technol 24:2668–2677. https://doi.org/10.1016/j.jmrt.2023.03.150
37. Gupta A, Kushwah KK, Mahobia SK, Soni P, Murty VVS (2019) Synthesis and characterization of TiO_2 nanoparticles for solar cell applications. Int J Innov Technol Explor Eng 8(9):2462–2465. https://doi.org/10.35940/ijitee.i8739.078919
38. Al-Bukhaiti WQ, Noman A, Wang H (2018) Emulsions: micro and nano-emulsions and their applications in industries-a mini-review. Int J Agric Innov Res 7(1). ISSN: 2319-1473
39. Zhang Y, Mei Q, Zhang Z (2024) Advancements in microemulsion-based fabrication of upconversion-mediated multifunctional materials. Front Photon 5:1–8. https://doi.org/10.3389/fphot.2024.1363223
40. Zuniga JP, Abdou M, Gupta SK, Mao Y (2018) Molten-salt synthesis of complex metal oxide nanoparticles. J Vis Exp 140:1–7. https://doi.org/10.3791/58482
41. Tran TQ, Zheng W, Tsilomelekis G (2019) Molten salt hydrates in the synthesis of TiO_2 flakes. ACS Omega 4(25):21302–21310. https://doi.org/10.1021/acsomega.9b02850
42. Lasfargues M, Bell A, Ding Y (2016) In situ production of titanium dioxide nanoparticles in molten salt phase for thermal energy storage and heat-transfer fluid applications. J Nanoparticle Res 18(6):1–11. https://doi.org/10.1007/s11051-016-3460-8
43. Darr JA, Zhang J, Makwana NM, Weng X (2017) Continuous hydrothermal synthesis of inorganic nanoparticles; applications and future directions. Chem Rev 117(11):1–248
44. Gan YX, Jayatissa AX, Yu Z, Chen X, Li M (2020) Hydrothermal synthesis of nanomaterials, vol 2020
45. Batool K et al (2024) Novel multinary nanocomposite of GO/$AlCrO_3$/SiO_2/Mn_3O_4/SnO_2: synthesis and electrochemical performance for energy storage system. J Taibah Univ Sci 18(1):2351619, pp 1–14. https://doi.org/10.1080/16583655.2024.2351619
46. Syuhada N, Yuliarto B (2018) Synthesis and characterization hierarchical three-dimensional TiO_2 structure via hydrothermal method. IOP Conf Ser Mater Sci Eng 367(1):012052. https://doi.org/10.1088/1757-899X/367/1/012052

47. Liu N, Chen X, Zhang J, Schwank JW (2014) A review on TiO_2-based nanotubes synthesized via hydrothermal method: formation mechanism, structure modification, and photocatalytic applications. Catal Today 225:34–51. https://doi.org/10.1016/j.cattod.2013.10.090
48. Bregadiolli BA, Fernandes SL, De Oliveira Graeff CF (2017) Easy and fast preparation of TiO_2-based nanostructures using microwave assisted hydrothermal synthesis. Mater Res 20(4):912–919. https://doi.org/10.1590/1980-5373-MR-2016-0684
49. Poolakkandy RR, Menamparambath MM (2020) Nanoscale advances soft-template-assisted synthesis: a promising approach for the fabrication of transition metal. Nanoscale Adv 2:5015–5045. https://doi.org/10.1039/D0NA00599A
50. Kumar R, Kumar S, Kumar M, Luthra G (2024) Approaches for the synthesis of nanomaterials, historical development and applications in different sectors: a review. Nanomed Nanotechnol Open Access: 1–10. https://doi.org/10.23880/nnoa-16000292
51. Oviroh PO, Akbarzadeh R, Pan D, Coetzee RAM, Jen TC (2019) New development of atomic layer deposition: processes, methods and applications. Sci Technol Adv Mater 20(1):465–496. https://doi.org/10.1080/14686996.2019.1599694
52. Smieszek A et al (2020) Titanium dioxide thin films obtained by atomic layer deposition promotes osteoblasts' viability and differentiation potential while inhibiting osteoclast activity—potential application for osteoporotic bone regeneration. Materials (Basel) 13(21):1–20. https://doi.org/10.3390/ma13214817
53. Ansari MZ et al (2024) Atomic layer deposition—a versatile toolbox for designing/engineering electrodes for advanced supercapacitors. Adv Sci 11(1):2303055
54. Barreca D, Fois E, Gasparotto A, Maccato C, Oriani M, Tabacchi G (2018) The early steps of molecule-to-material conversion in chemical vapor deposition (CVD): a case study. Molecules 26(7):1–22
55. Wang H et al (2022) Large-scale chemical vapor deposition synthesis of graphene nanoribbions/carbon nanotubes composite for enhanced membrane capacitive deionization. J Electroanal Chem 904:115907. https://doi.org/10.1016/J.JELECHEM.2021.115907
56. Ura DP, Stachewicz U (2022) The significance of electrical polarity in electrospinning: a nanoscale approach for the enhancement of the polymer fibers' properties. Macromol Mater Eng 307(5):1–23. https://doi.org/10.1002/mame.202100843
57. Zhou W et al (2021) Environmentally friendly waterborne polyurethane nanofibrous membranes by emulsion electrospinning for waterproof and breathable textiles. Chem Eng J 427:130925. https://doi.org/10.1016/j.cej.2021.130925
58. Keskinen H (2007) Synthesis of nanoparticles and preparation of deposits by liquid flame spray synthesis of nanoparticles and preparation of deposits by liquid flame spray
59. Falcony C, Aguilar-Frutis MA, García-Hipólito M (2018) Spray pyrolysis technique; high-K dielectric films and luminescent materials: a review. Micromachines 9(8):1–33. https://doi.org/10.3390/mi9080414
60. Widiyandari H, Gunawan V, Widyanto SA, Purwanto A (2017) Synthesis of titanium dioxide (TiO_2) fine particle by flame spray pyrolysis (FSP) method using liquid petroleum gas (LPG) as fuel. Reaktor 17(4):226–229. A credited: SK no.: 60/E/KPT/2017 website: http://ejournal.undip.ac.id/index.php/reaktor/
61. Bertus LM, Enesca A, Duta A (2012) Influence of spray pyrolysis deposition parameters on the optoelectronic properties of WO_3 thin films. Thin Solid Films 520(13):4282–4290. https://doi.org/10.1016/j.tsf.2012.02.052
62. Nyamukamba P, Okoh O, Mungondori H, Taziwa R, Zinya S (2018) Synthetic methods for titanium dioxide nanoparticles: a review. Titan Dioxide Mater Sustain Environ 338(5):113485. https://doi.org/10.5772/intechopen.75425
63. Gomaa MM, Abdel-wahed MS (2023) RSC advances synthesis of Sb_2S_3 nanosphere layer by chemical bath deposition for the photocatalytic degradation of methylene blue dye. RSC Adv 13:22054–22060. https://doi.org/10.1039/D3RA02062B
64. Zyoud SH, Ahmed NM, Lahewil AS, Omar AF (2022) Micro spot ZnO nanotubes using laser assisted chemical bath deposition: a low-cost approach to UV photodetector fabrication. Sens Actuators A Phys. https://www.researchgate.net/publication/359187151_micro_spot_zno_

nanotubes_using_laser_assisted_chemical_bath_deposition_a_low-cost_approach_to_uv_photodetector_fabrication. Accessed 19 June 2024

65. Zyoud SH et al (2023) Novel laser-assisted chemical bath synthesis of pure and silver-doped zinc oxide nanoparticles with improved antimicrobial and photocatalytic properties. Catalysts 13(900):1–23
66. Zhu Y, Chen F (2014) Microwave-assisted preparation of inorganic nanostructures in liquid phase
67. Nain S, Singh R, Ravichandran S (2019) Importance of microwave heating in organic synthesis. Adv J Chem Sect A 2(2):94–104. https://doi.org/10.29088/SAMI/AJCA.2019.2.94104
68. Wu L, Shen J, Jin B (2024) Microwave-assisted synthesis of few-layer $Ti_3C_2T_x$ loaded with $Ni_{0.5}Co_{0.5}Se_2$ nanospheres for high-performance supercapacitors. Materials (Basel) 17(10):2292
69. Salvatore KL et al (2023) Microwave-assisted synthesis of iron sulfide motifs for electrochemical applications Microwave-assisted synthesis of iron sulfide motifs for electrochemical applications. J Phys Mater 6(2023):024005
70. Karim N et al (2023) Green synthesis of nanolipo-fibersomes using Nutriose® FB 06 for delphinidin-3-O-sambubioside delivery: characterization, physicochemical properties, and application. Int J Biol Macromol 247:125839. https://doi.org/10.1016/j.ijbiomac.2023.125839
71. Xu Y et al (2023) Synthesis and potential applications of cyclodextrin-based metal–organic frameworks: a review, vol 21, no 1. Springer International Publishing. https://doi.org/10.1007/s10311-022-01509-7
72. Behera M, Behera PR, Bhuyan PP, Singh L, Pradhan B (2023) Algal nanoparticles and their antibacterial activity: current research status and future prospectives. Drugs Drug Candidates 2(3):554–570. https://doi.org/10.3390/ddc2030029
73. Michael A, Singh A, Roy A, Islam MR (2022) Fungal- and algal-derived synthesis of various nanoparticles and their applications. Bioinorg Chem Appl 2022. https://doi.org/10.1155/2022/3142674
74. Thatyana M, Dube NP, Kemboi D, Manicum ALE, Mokgalaka-Fleischmann NS, Tembu JV (2023) Advances in phytonanotechnology: a plant-mediated green synthesis of metal nanoparticles using phyllanthus plant extracts and their antimicrobial and anticancer applications. Nanomaterials 13(19):2616. https://doi.org/10.3390/nano13192616
75. Monga Y et al (2020) Sustainable synthesis of nanoscale zerovalent iron particles for environmental remediation. Chemsuschem 13(13):3288–3305. https://doi.org/10.1002/cssc.202000290
76. Omran BA, Omran BA (2020) Biosynthesized nanomaterials via processing of different plant parts (phytonanotechnology) and biovalorization of agro-industrial wastes to nano-sized valuable products. Nanotechnol Life Sci: 145–184. https://doi.org/10.1007/978-3-030-46071-6_5
77. Nath PC et al (2023) Biogeneration of valuable nanomaterials from agro-wastes: a comprehensive review. Agronomy 13(2):1–37. https://doi.org/10.3390/agronomy13020561
78. Osman AI, Zhang Y, Farghali M, Rashwan AK, Eltaweil AS (2024) Synthesis of green nanoparticles for energy, biomedical, environmental, agricultural, and food applications: a review, vol 22, no 2. Springer International Publishing. https://doi.org/10.1007/s10311-023-01682-3
79. Abuzeid et al (2023) Green synthesis of nanoparticles and their energy storage, environmental, and biomedical applications. Storage Energy Appl Biomed 13(1576):1–51

Chapter 6
Advanced Synthesis Strategies for Enhanced Energy Storage Performance

Jayanti Mishra, Arabinda Baruah, and Jyotirekha Dutta

Abstract This chapter discusses the synthesis of the popular materials used in energy storage devices. The popular materials used in such devices are layered double hydroxides, graphene and polymer-based nanostructures, perovskites, MOFs and COFs, Mxene and MBenes, hydrogels, waste biomass-derived materials, nanocomposites, etc. for energy storage applications. These devices include supercapacitors, mechanical and thermal energy storage devices and various batteries working upon Na-ion, K-ion, Zn-ion, Al-ion and Li-ion. In this chapter, various synthesis methods like co-precipitation, solvothermal, hydrothermal, electrospinning, self-assembly, hot injection, sonochemical method, etc. are discussed for the synthesis of suitable materials for application in energy storage devices.

Keywords Advanced synthesis strategies · Enhanced energy storage performance

6.1 Layered Double Hydroxides (LDHs) Based Hybrid Nanostructures for Energy Storage Applications

The globe is currently searching for different non-conventional energy options due to the rapid depletion of fossil fuels and growing environmental concerns. The intermittent nature of non-conventional energy sources like solar, wind, tidal, geothermal, etc. means that electrochemical energy storage systems are crucial to resolving the issue of energy crisis. Among various nanostructured materials for energy storage devices,

J. Mishra (✉)
Department of Chemistry, New Horizon College of Engineering, Bengaluru, Karnataka 560103, India
e-mail: jayanti.mishra2001@gmail.com

A. Baruah
Department of Chemistry, Gauhati University, Guwahati, Assam 781014, India

J. Dutta
Department of Chemistry, Indian Institute of Technology Hyderabad, Sangareddy, Telangana 502284, India

R. Naik et al. (eds.), *Advanced Nanomaterials for Energy Storage Devices*, Nanostructure Science and Technology, https://doi.org/10.1007/978-3-031-74730-4_6

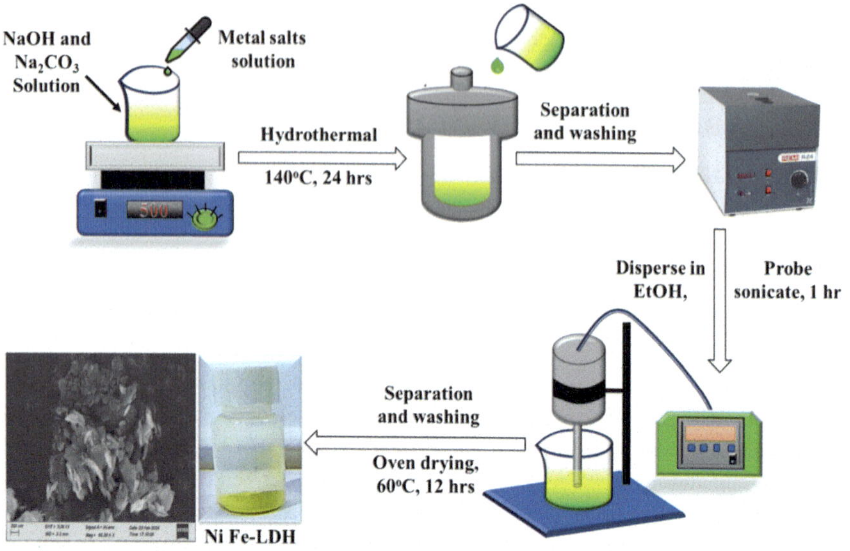

Fig. 6.1 Synthesis of Fe–Ni LDH nanoparticles via hydrothermal method [1]

Layered Double Hydroxides are found to be highly efficient. Layered double hydroxides (LDHs) have the chemical formula of $[M_{1-x}{}^{2+} M_x{}^{3+}(OH)_2]^{x+} [A^{n-}{}_{x/n}\cdot{}_yH_2O]^{x-}$, where M^{2+} and M^{3+} are the divalent and trivalent metal cations located in the brucite-like host layers and A^n is the charge compensating anion in the interlayer region. These particular layered nanostructured materials are distinct in kind and can be easily synthesized via solvothermal route (Fig. 6.1).

The following characteristics of LDHs' structure make them attractive materials for use in electrochemical energy conversion and storage systems: (a) large surface area accessible to ions; (b) sufficient rate of ionic transport; (c) substantial capacity; (d) remarkable mechanical properties; and (e) highly adjustable electric and thermal conductivity. Because batteries are utilized on-demand and have intermediate energy and power densities, their applications market is larger than that of any other electrochemical energy storage devices (fuel cell, capacitor, supercapacitor, battery). Rechargeable Li-ion battery demand has been rising dramatically over the past 20 years. This exciting technology now occupies a significant piece of the energy market. Li is the lightest metal (6.94 g/mole) with the lowest reduction potential of − 3.04 V versus SHE which can provide a high theoretical capacity of 3861 mAh/g, highest possible cell potential, high energy density, and long cycle life.

Since Li metal is highly reactive, it forms dendrites until a solid electrolyte layer (SEI) known as stable passivation forms on the Li electrode's surface. This problem can be solved by using the layered structure of graphite and the "rocking chair" principle, in which the Li^+ ion is intercalated and de-intercalated from the layers of the material by graphite acting as a host material. Propylene carbonate (PC) solvent's co-intercalation property restricts the use of graphite as the anode. The first commercially available anode material for Li-ion batteries that does not collapse structurally was petroleum coke. But the main constraint is the petroleum coke's limited ability to store energy.

Therefore, numerous advanced nanomaterials have been developed to overcome this issue. For instance, Huang and colleagues presented an easy method to create a Ni–Co layered double hydroxide (LDH) on conductive porous micro-structured stainless-steel yarn (SSY@CPMs) electrode through hydrothermal growth. This electrode has mechanical stability, hierarchical structures, and exceptional capacitive properties because of its high porosity and improved dispersive adhesion. They also put together a cutting-edge, all-solid-state, asymmetric supercapacitor (ASC) based on the Ni–Co–SSY@CPMs electrode that has outstanding mechanical and electrochemical performance. The gadget demonstrated an exceptional cycle capability while displaying a maximum energy density of 22.52 μW h cm^{-2} and a power density of 7.43 mW cm^{-2} [2]. Li et al. reported Nickel aluminium layered double hydroxide (Ni–Al LDH) with nitrate in its interlayer as anode material for Li-ion batteries. The effect of binder, sodium alginate (SA), and polyvinylidene difluoride (PVDF), and of the potential range (i.e., 0.01–3.0 and 0.4–3.0 V vs. Li^+/Li), on the performance of the anode material was investigated in 1 M $LiPF_6$ in EC/DMC electrolyte [3].

Compared to numerous electrochemical energy storage devices, supercapacitors have high-power density, long cycle life, fast charge–discharge process, and unlimited performance. But the biggest problem that limits the application market of supercapacitors is low energy density. Many researchers already reported Ni–Co LDHs, Co–Al LDHs, Co–Fe LDHs, Ni–Al LDHs, Ni–Co–Al LDHs, Co–Mn LDHs, Ni–Mn LDHs, Ni–V LDHs, etc. as excellent electrode materials for supercapacitors. Tyagi et al. prepared Ni–V-LDH nanosheets by hydrothermal method for applications in high-power density supercapacitors. By varying the Ni:V ratios, they reported $Ni_{0.80}V_{0.20}$-LDH showed better electrochemical performance, a high specific capacity of 711 C/g at a current density of 1 A/g. The hybrid supercapacitor device (HSC), is prepared by assembling $Ni_{0.80}V_{0.20}$-LDH nanosheets as a positive electrode with commercial activated carbon as the negative electrode at a working voltage of 1.5 V. The HSC device showed a specific capacitance of 98 F/g at a current density of 1 A/g, and a higher energy density of 30.6 Wh/kg at a power density of 0.78 kW/kg [4]. Zhou et al. synthesized an ultrathin, porous, sulfurated NiV-LDH, (NiV-S) through an ion exchange reaction (S^{2-} replaces OH^-). The reported NiV–S showed a higher specific capacitance of 2270.4 F/g at 2 A/g, cycling stability of 91.9% retention after 10,000 cycles [5]. Zhou et al. synthesized NiMn LDHs/halloysite electrode materials by deposition of NiMn LDHs and poly(3,4-ethylenedioxythiophene) on halloysite nanotubes via growth and polymerization. The reported NiMn LDHs/halloysite electrode materials showed a specific capacitance of 1665.6 F/g at 1 A/g [6]. The two main drawbacks of transition metal LDHs during cycling are their slow electrolyte penetration and limited electrical conductivity. Research is currently concentrated on creating composites with highly conductive nanoparticles in order to address these problems with LDHs.

6.2 Graphene-Based Nanostructures for Energy Storage Applications

Graphene, a single sheet of carbon atoms organized in a two-dimensional honeycomb lattice, has attracted a lot of interest for its potential applications in diverse domains of research [7]. Because of its desirable qualities, it is being used in fabricating energy storage devices too. Graphene's high specific surface area (~ 2630 m^2/g) allows for a greater electrode–electrolyte interface, increasing the capacitance. It exhibits excellent electrical conductivity, enhancing charge and discharge rates. Moreover, it offers high cycle stability, maintaining performance over many charge/discharge cycles. People have widely used various graphene-based materials, such as graphene oxide, reduced graphene oxide, hybrid nanocomposites and functionalized graphene for designing energy storage devices.

Graphene oxide can be easily prepared using Hummer's method or modified Hummer's method where, H_2SO_4 and $NaNO_3$ are added to graphene at 0 °C followed by addition of $KMnO_4$ at temperature below 20 °C, followed by addition of water at 98 °C and addition of H_2O_2. After vacuum drying at 40 °C and exfoliation graphene oxide is obtained. Apart from this, Brodie's method, Staudenmaier method, Hoffman's method and Tour's method are also being popularly used for graphene oxide synthesis.

In Li-ion batteries, graphene can be used as an anode material or combined with silicon to improve capacity and lifespan. As a conductive additive in electrodes, graphene can enhance electrical conductivity and mechanical stability. In addition, it facilitates faster electron and ion transport, leading to quicker charging times. Doping graphene with nitrogen or boron, its energy storage capacity can be further improved. For example, Wu et al. reported that N-doped graphene and B-doped graphene exhibit high reversible capacity of 1043 and 1540 mA h g^{-1} at a low rate of 50 mA g^{-1}, respectively. Moreover, a high capacity of 199 and 235 mA h g^{-1} was observed for the N-doped graphene and B-doped graphene at a high rate of 25 A g^{-1} (about 30 s to fully charge), respectively. The disordered surface morphology, heteroatomic defects, improved surface wettability, increased interlayer distance, improved electrical conductivity, and thermal stability of N- and B-doped graphene contribute to its high-rate performance. These properties are advantageous for electron transport, ultrafast Li^+ diffusion, and rapid surface Li^+ absorption [8].

In lithium-sulphur batteries it is being used to help in immobilizing sulphur and mitigate the polysulphide shuttle effect, improving efficiency and lifespan. The overall conductivity of sulphur-based cathodes is also enhanced due to the use of graphene. By acting as supporting material to catalysts like platinum in fuel cells, graphene lowers the quantity of precious metal needed while boosting efficiency and cutting costs. It also offers corrosion resistance and mechanical stability. Graphene can replace materials like indium tin oxide (ITO) in transparent conductive sheets used in solar cells. Moreover, it functions as an effective layer for charge transport, enhancing the efficiency of perovskite solar cells and organic photovoltaics.

Technologies for energy storage could be revolutionized by graphene. The performance, endurance, and efficiency of energy storage devices might all be significantly improved with more research and development in this field.

6.3 Polymer-Based Nanostructures for Energy Storage Applications

Polymers are a class of material consisting of lengthy chains of monomers. They can be produced chemically by a variety of processes, or they can be found naturally in substances like cellulose, proteins, and DNA. Polymer materials are being investigated very often for energy storage applications because of their superior qualities, which include flexibility, light weight, and formability. Several polymers have been studied and evaluated in recent years for diverse uses including energy storage. Polymer electrolytes (such as, polyethylene oxide, polyvinyl alcohol, polymethyl methacrylate, and polycaprolactone are particularly important among them as they are being used in solid-state batteries and supercapacitors. They offer several advantages, including, high ionic conductivity, mechanical stability and flexibility. Conducting polymers like polyaniline (PANI), polypyrrole (PPy), and polythiophene (PTh) are used in supercapacitors and batteries as electrodes due to their high electrical conductivity and electrochemical stability. They can also enhance the energy density and power density of devices.

To synthesize such conducting polymers, there are several methods reported in the literature [9]. Among them, one of the most common methods for synthesizing polyaniline, is chemical oxidation, which involves mixing a monomer precursor with an oxidizing agent in the presence of acid. Electrochemical polymerization is the go-to method for synthesizing highly conductive polypyrrole, which involves the anodic oxidation of a monomer precursor in the presence of suitable electrolytes on an inert metal surface. Vapour phase synthesis method involves depositing the monomer onto a substrate and then oxidizing it to create a film of conductive polymer. Other common methods like, hydrothermal and solvothermal synthesis involve heating the monomer in a solvent to create a crystalline film of conductive polymer. In template-assisted synthesis, a template is used to control the shape of the conductive polymer as it forms. Electrospinning method involves using an electric field to draw out the conductive polymer into fibres. In self-assembly method, the conductive polymer is allowed to form spontaneously into a specific shape or structure. Apart from these, there are other methods, like, plasma polymerization, where, plasma is used to initiate the polymerization reaction.

6.4 Perovskite for Energy Storage Applications

Halide perovskites have good ionic, electronic, optoelectronic, ferroelectric, piezoelectric and optoionic properties which make them suitable for energy harvesting and energy storage devices. Popoola et al. used halide-perovskite ($Cu_3Bi_2I_9$) material in fabricating a Cu-perovskite photoactive electrode as supercapacitor for energy harvesting and electrochemical energy storage. They synthesized HPvA gel electrolyte and sandwiched it between two Cu-perovskite photoactive electrodes for 127% energy density augmentation with scan rate of 0.01 V/s. 0.5897 g of BiI_3 and 0.5713 g of CuI (3:1) were added in 1 mL of DMSO solvent to get $Cu_3Bi_2I_9$. It was stirred at 1200 rpm at 90 °C, for 8 h. The product was stored in nitrogen-filled glove box after that [10]. Ippili et al. designed methylammonium tin halide (MA_2SnX_6; X = Cl, Br, and I), a double perovskite thin film for highly efficient and stable self-charging power units (SPUs). They prepared MA_2SnCl_6-based piezoelectric nanogenerator (PENG) with a lithium metal battery (LMB) with MA2SnCl6-based thin film as the cathode, which shows high output power density of 7.33 μW cm^{-2} and good mechanical durability [11]. MA_2SnX_6 thin films were prepared by a simple one-step spin-coating technique under ambient air conditions. Three different MA_2SnX_6 (1 M) precursor solutions were prepared by dissolving equimolar ratio of MAX and SnX_2 materials in 1 mL DMF solvent which was stirred continuously on a hot plate at 90 °C for 2 h. Popoola et al. fabricated the perovskite electrochemical supercapacitors by using the quasi-solid-state electrolytes sandwiched between two symmetric perovskite electrodes with better areal capacitance, high power density. 406 mg of PbI_2 and 140 mg of CH_3NH_3I are dissolved in 1 mL of DMF solvent which was stirred at 70 °C for 8 h to get the perovskite precursor material i.e., active methylammonium Lead Iodide ($MAPbI_3$), which was filtered and stored in glove box filled with nitrogen gas [12]. Luminescent nanocrystals are formed by hot injection method with the incorporation of various ligands. The precursor solution is quickly injected into another hot stirring precursor solution which is followed by the reaction quenching in an ice bath. Injection speed, choice of ligands and reaction duration determine the size and shape of the nanocrystal products. Moderate levels of strategic vacancies make them a better prospect for energy storage material [13].

6.5 MOFs and COFs for Energy Storage Applications

Metal–organic framework (MOF) derivatives are proven to be very efficient active material of energy storage devices because of their large surface area, abundant redox states and excellent pore structure. Ammonia fluoride was used to control morphology of zeolitic imidazolate framework 67 (ZIF-67) and supply $NH_4{}^+$ and F^- for better charge transfer. Cheng et al. sulfurized ammonium-decorated cobalt fluoride (S-NH_4-CoF) using ZIF-67 derivative and sulphur powder by one-step annealing process at various temperatures. The S-NH4-CoF electrode which is has round particle-assembled sphere morphology, prepared at 400 °C exhibits specific capacitance (CF) of 462.1 F g^{-1} at 20 mV s^{-1} with high surface area and large pore volume, and suitable ratio of cobalt fluoride and cobalt sulphide and maximum energy density of 30.47 Wh kg^{-1} at 800 W kg^{-1}. One solution which contains 3 mmol of $Co(NO_3)_2{\cdot}6H_2O$ and 13.5 mmol of NH_4F in 30 ml of methanol is mixed with the other solution which contains 12 mmol of 2-methylimidazole in 10 ml of methanol and stirred at room temperature for 24 h. The precipitates were filtered and washed in methanol and dried at 60 °C overnight. The sulfurization was done by annealing NH_4-CoF and S powder (2:1) in a tube furnace at 300, 400, 500 and 600 °C in the N_2 atmosphere for 2 h [14].

6.6 Mxene and MBenes for Energy Storage Applications

2D transition metal carbides and nitrides are termed as MXenes and 2D transition metal borides are termed as MBenes. They have high conductivity, flexibility and mechanical properties which are making suitable for applications in energy storage devices like batteries and supercapacitors. Li et al. prepared MoB, MgB_2, and ZrB_2 for cathode material in Zn-ion batteries. They mixed MoAlB powder (1 g) with NaOH solution (25 wt%, 35 mL) and soaked it in water for 8 h. It was then transferred to autoclave for heating at 150 °C for 24 h. After cooling, the product was filtered and washed with 1 M NaOH and water and then dried at 60 °C under vacuum. For MgB_2, they dispersed original MgB_2 (0.18 g) powder with PEG4000 (0.9 g) in 240 ml deionized water and then CH_3COOH (1080 μl) and H_2O_2 (30%, 18 mL) followed by ultrasound. The solution was washed with deionized water and dried at 60 °C under vacuum. For ZrB_2, original ZrB_2 powder (0.17 g) was mixed with F68 (0.45 g) and CH_3COOH (385 mg) in ethylene glycol followed by addition of H_2O_2 (30%, 3 mL). This solution was transferred to microwave digestion tank followed by microwave chemical reactor and heated for 2 h at 160 °C and 400 W. After that, it was cooled, washed and dried at 60 °C [15].

6.7 Hydrogels for Energy Storage Applications

Hydrogels are very promising materials for electrolytes as they contain high water amount and flexible mechanical properties, porous structure but it is important for them to be adaptable towards environment to save themselves from failure. They should be able to avoid freezing or dehydration in cold and arid environments for sustainable electrochemical performance [16]. Tong et al. developed a cross-linked cellulose-based hydrogel electrolyte having covalent and hydrogen bonding which facilitates the ionic conductivity of Zn^{2+} at room temperature to suppress the formation of Zn dendrites in Zn-ion energy storage devices [17]. They have taken the different weights of cellulose particles to alkaline urea solution to make up the total mass 10 g. It was stirred vigorously at − 12 °C to get a clear solution followed by addition of (±)-Epichlorohydrin (1:1 to cellulose solution) which was degassed and placed at 5 °C for 12 h for chemical cross-linking. It was washed followed by physical cross-linking. Liu et al. synthesized an epoxy resin/ hydrogen bicontinuous phase structure using hydrophobic curing agent and the substitution of hydrogel for water. They mixed KOH (31.0 g in 50 mL water at 90 °C) solution and PVA solution (20.0 g in 200 mL water at 95 °C) to form the PVA/KOH hydrogel by mixing at 90 °C for 30 min. This gel was mixed with different ratios of epoxy to create various solid electrolytes [18].

6.8 Waste Biomass-Derived Materials for Energy Storage Applications

Biomass is a major and cheap source of activated carbon which can be used in energy storage devices. Also, this will be eco-friendly which will keep environment along with producing energy. Hegde et al. have developed *Mangifera indica* (mango) leaf waste-derived activated carbon and used it as electrode material for supercapacitors. They sundried and washed with pure water followed by ethanol and dried again in the oven. It was then carbonized by using $FeCl_3$ at 100 °C followed by hot air oven treatment at 110 °C for 24–48 h followed by 1.5 h treatment at 400 °C under Ar gas environment and activated by KOH (10 g) in water with stirring at hotplate followed by drying in hot air oven at ~ 110 °C for 24–48 h which was pyrolyzed after that at various temperature in Ar gas. This resulted in an increase in its surface area and pore structure [19]. Li et al. designed the nitrogen-doped and nitrogen-boron co-doped carbon electrode using cherry blossoms (*Prunus yedoensis*) as biomass source. This nitrogen- and boron-doped porous carbon has better cross-linking and pore structure. This material has strong adsorption capacity for K ions. Its structure helps in charge migration of neighbouring carbon atoms which results in enhancement of diffusion of the electrolyte ions. They prepared the porous KOH-activated cherry blossom derived carbon by cleaning and blast drying cherry blossom petals at 110 °C overnight followed by ball milling treatment to get the powder. This powder was heated up to 550 °C in tubular furnace under nitrogen at 5 °C/min for 1 h followed by grinding to get fine powder. It was then activated with KOH (3 times), stirred and blast dried at 110 °C overnight. This powder was heated in tube furnace at 5 °C/min under nitrogen for 2 h at 750 °C. It was acid washed followed by water wash and then dried under vacuum at 65 °C for 5 h. For doping of heteroatom, hydrothermal reaction was set up. This KOH-activated cherry blossom derived carbon was mixed with boric acid 1:4) in presence of distilled water and mixed under ultrasonication. Then, it was kept in autoclave at 180 °C for 14 h. It was cooled and washed and vacuum dried at 65 °C for 5 h. After boron doping, it was mixed with ammonia and distilled water under hydrothermal to get boron and nitrogen-doped carbon material [20]. Similarly, waste material from *Cocos nucifera* (coconut) and Algal biomass are also proven to be great materials for energy storage devices [21, 22].

6.9 Nanocomposites for Energy Storage Applications

Kamran et al. prepared bifunctional chromium doped Cadmium Sulphide/Zinc Telluride (Cr–CdS/ZnTe) nanocomposites for better performance in energy storage devices. Its morphology is nanoflake-like. They prepared it by sonochemical method. At first, they prepared CdS, Cr–CdS and ZnTe nanopowder by co-precipitation method. For CdS, they mixed cadmium acetate (0.1 M) and polyvinylpyrolidone followed by dropwise addition of sodium sulphide (0.1 M). It was stirred and precipitates were acquired. For Cr-doped CdS, different amount of PVP were added followed by addition of different amount of chromium chloride. The precipitate was collected, washed and dried at 8 °C in oven for 6 h. Similarly, ZnTe nanoparticles are prepared by mixing zinc acetate (0.25 M) and PVP (0.5 g) followed by dropwise addition of sodium telluride (Na_2TeO_3) (0.25 M) solution. ZnTe nanopowder was obtained and cleaned. For preparation of Cr–CdS/ZnTe nanocomposites, Cr–CdS and ZnTe nanopowder were mixed and sonicated for 6 h in ultrasonic chamber. The mixture was filtered, washed and dried at 80 °C for 6 h followed by calcinations in muffle furnace at 700 °C for 4 h [23]. Rafeeq et al. synthesized Nd_2O_3/Sm_2O_3 nanocomposite as electrode material in energy storage devices [24]. Chen et al. prepared sandwich structured relaxor ferroelectric nanocomposite material which is incorporated with core–shell fillers for amazing energy storage applications [25]. Md. Islam et al. prepared a MoS_2/MnO_2 nanocomposite electrode which has 3D nanoflower/1D nanorod architecture by hydrothermal process. It is combination of 1D MnO_2 nanorod (for structure backbone) and 3D MoS_2 nanoflower (for reactive sites). It facilitates fast electron and ion movement which increases charge collection capacity for energy storage devices. MoS_2 was prepared by mixing sodium molybdate dehydrate (0.14 M) and thiourea (0.65 M) in water. It was then kept in teflon-lined autoclave at 200 °C for 24 h. The precipitate was collected and dried at 70 °C. MnO_2 was prepared by mixing $MnSO_4.H_2O$ (5.6878 g) and $KMnO_4$ (2.3657 g) for 30 min. It was then kept in autoclave for 12 h in oven at 140 °C. The precipitate was collected and dried at 80 °C. MnO_2 incorporated MoS_2 was prepared by taking MnO_2 nanorod in water and sonicating it for 60 min. Then sodium molybdate dehydrate (0.14 M) and thiourea (0.65 M) were dissolved in deionized water. This solution was transferred to sonicated MnO_2 solution (50 mL) followed by stirring for 60 min. The solution was kept in autoclave at 200 °C for 24 h. The precipitate was collected at dried at 80 °C [26].

6.10 Conclusion

In this chapter, some of the literature reports have been discussed where various methods have been used to synthesize materials like MOFs, COFs, hydrogels, LDHs, perovskites, polymers, graphene, MXenes and MBenes, etc. which are used in energy storage devices. Their hybrids like MOFs/graphene oxide, MOFs/ MXenes, metal-doped perovskites, MOF/ metal oxide, MXene/polymer, etc. are also evolving as better material for energy storage applications. The synthesis method ensures the rapid movement of ions across electrolyte for better performance. The common and popular methods used were hydrothermal method, co-precipitation method, modified Hummer's method for graphene oxide, ultrasonication, microwave, etc. which are used to synthesize above materials. This chapter aimed to bring the materials together which are suitable for energy storage applications.

References

1. Baruah A, Mondal S, Sahoo L, Gautam UK (2019) Ni–Fe-layered double hydroxide/N-doped graphene oxide nanocomposite for the highly efficient removal of Pb(II) and Cd(II) ions from water. J Sol State Chem 280:120963. https://doi.org/10.1016/j.jssc.2019.120963
2. Huang J, Xu B, So MY, Liu X, Gao Y (2024) A hierarchical layered double hydroxide electrode with surface porous microstructured fibers for flexible and wearable energy storage. J Mater Chem A 12(12):7187–7198. https://doi.org/10.1039/d3ta06281c
3. Li X, Fortunato M, Cardinale AM, Sarapulova A, Njel C, Dsoke S (2021) Electrochemical study on nickel aluminum layered double hydroxides as high-performance electrode material for lithium-ion batteries based on sodium alginate binder. J Sol State Electrochem 26(1):49–61. https://doi.org/10.1007/s10008-021-05011-y
4. Tyagi A, Joshi MC, Shah A, Thakur VK, Gupta RK (2019) Hydrothermally tailored three-dimensional Ni–V layered double hydroxide nanosheets as high-performance hybrid supercapacitor applications. ACS Omega 4(2):3257–3267. https://doi.org/10.1021/acsomega.8b03618
5. Zhao Y, Ma H, Huang S, Zhang X, Xia M, Tang Y, Ma Z-F (2016) Monolayer nickel cobalt hydroxyl carbonate for high performance all-solid-state asymmetric supercapacitors. ACS Appl Mater Interfaces 8(35):22997–22995. https://doi.org/10.1021/acsami.6b05496
6. Zhou J, Dai S, Li Y, Han F, Yuan Y, Tang J, Tang W (2018) Earth-abundant nanotubes with layered assembly for battery-type supercapacitors. Chem Eng J 350:835–43. https://doi.org/10.1016/j.cej.2018.06.030
7. Baruah A, Newar R, Das S, Kalita N, Nath M, Ghosh P, Chinnam S, Sarma H, Narayan M (2024) Biomedical applications of graphene-based nanomaterials: recent progress, challenges, and prospects in highly sensitive biosensors. Discover Nano 19(1). https://doi.org/10.1186/s11671-024-04032-6
8. Wu Z-S, Ren W, Xu L, Li F, Cheng H-M (2011) Doped graphene sheets as anode materials with superhigh rate and large capacity for lithium ion batteries. ACS Nano 5(7):5463–71. https://doi.org/10.1021/nn2006249
9. Namsheer K, Rout CS (2021) Conducting polymers: a comprehensive review on recent advances in synthesis, properties and applications. RSC Adv 11(10):5659–97. https://doi.org/10.1039/d0ra07800j

10. Popoola IK, Gondal MA, Popoola A, Oloore LE, Younas M (2023) Inorganic perovskite photo-assisted supercapacitor for single device energy harvesting and storage applications. J Energy Storage 73:108828. https://doi.org/10.1016/j.est.2023.108828
11. Ippili S, Kim JH, Jella V, Behera S, Vuong V-H, Jung J-S, Cho Y et al (2023) Halide double perovskite-based efficient mechanical energy harvester and storage devices for self-charging power unit. Nano Energy 107:108148. https://doi.org/10.1016/j.nanoen.2022.108148
12. Popoola I, Gondal M, Oloore L, Popoola AJ, AlGhamdi J (2020) Fabrication of organometallic halide perovskite electrochemical supercapacitors utilizing quasi-solid-state electrolytes for energy storage devices. Electrochimica Acta 332:135536. https://doi.org/10.1016/j.electacta.2019.135536
13. Choon SL, Lim HN (2024) Impact of vacancies in halide perovskites for batteries and supercapacitors. Mater Today Energy 43:101577. https://doi.org/10.1016/j.mtener.2024.101577
14. Cheng T-M, Chu H-Y, Cao Y-C, Kubendhiran S, Lee P-Y, Yougbaré S, Lin L-Y, Wu Y-F (2022) Design of novel cobalt-based perovskite derivatives using sulfurization as active material of energy storage devices. J Energy Storage 55:105653. https://doi.org/10.1016/j.est.2022.105653
15. Li Z, Cheng Y, Wang Y, Cheng J, Qiu J, Shi Y, Wang R (2024) Exploring the potential of MBenes in energy storage. In: Colloids and surfaces a physicochemical and engineering aspects, 134317. https://doi.org/10.1016/j.colsurfa.2024.134317
16. Chen Y, He S, Rong Q (2023) Recent progress in environment-adaptable hydrogel electrolytes for flexible energy storage devices. J Energy Storage 73:109023. https://doi.org/10.1016/j.est.2023.109023
17. Tong M, Kuang S, Wang Q, Li X, Yu H, Zeng S, Yu X (2024) Dual cross-linked cellulose-based hydrogel for dendrites-inhibited flexible zinc-ion energy storage devices with ultra-long cycles and high energy density. Carbohydrate Polym: 122444. https://doi.org/10.1016/j.carbpol.2024.122444
18. Liu H, Zhang S, Ma Z, Chen H, Chen F, Wang K, Zhang C, Li H (2024) Fabrication of novel hydrogel/epoxy resin composites with soft-hard combined bi-continuous aqueous load-bearing electrolytes and application in structural energy storage devices. Chem Eng J 486:150295. https://doi.org/10.1016/j.cej.2024.150295
19. Hegde SS, Bhat BR (2024) Sustainable energy storage: *Mangifera indica* leaf waste-derived activated carbon for long-life, high-performance supercapacitors. RSC Adv 14(12):8028–8038. https://doi.org/10.1039/d3ra08910j
20. Li P-H, Wu W-J (2024) Co-doping mechanism of biomass-derived nitrogen-boron porous carbon and its applications in energy storage and environmental purification. J Alloys Compounds: 175098. https://doi.org/10.1016/j.jallcom.2024.175098
21. Hegde SS, Bhat BR (2024) Impact of electrolyte concentration on electrochemical performance of *Cocos nucifera* waste-derived high-surface carbon for green energy storage. Fuel 371:131999. https://doi.org/10.1016/j.fuel.2024.131999
22. Cha J, Lim C-H, Lee J, Lim J-K, Kim M, Park W-K, Sung C, Kim H-K, Min K (2024) Algal biomass-based zero-waste biorefinery for producing optically pure (R)-γ-valerolactone and carbonaceous electrodes applicable for energy storage devices. Chem Eng J: 151713. https://doi.org/10.1016/j.cej.2024.151713
23. Kamran MA, Raza M, Ullah S, Alarifi SS, Aziz MH (2024) Facile synthesis of chromium doped cadmium sulfide/zinc telluride nanocomposites for enhanced electrochemical energy storage and photocatalytic applications. Mater Chem Phys 322:129531. https://doi.org/10.1016/j.matchemphys.2024.129531
24. Rafeeq M, Shah SIA, Jabbour K, Ahmad S, Abdullah M, Alshgari RA, Mohammad S, Ehsan MF, Yasmeen G, Ashiq MN (2024) Facile fabrication of Nd_2O_3/Sm_2O_3 nanocomposite as a robust electrode material for energy storage applications. J Energy Storage 88:111580. https://doi.org/10.1016/j.est.2024.111580
25. Chen Y, Xue Y, Shi J, Zhou J, Bai K, Li Y, Ma R et al (2024) Sandwich-structured relaxor ferroelectric nanocomposite incorporated with core-shell fillers for outstanding-energy-storage capacitor application. J Alloys Compounds: 175642. https://doi.org/10.1016/j.jallcom.2024.175642

26. Islam MR, Rahaman M, Billah MM, Islam MR (2024) Hydrothermal synthesis of MoS_2/MnO_2 nanocomposite: a unique 3D-nanoflower/1D-nanorod structure for high-performance energy storage applications. Mater Adv 5(12):5307–5321. https://doi.org/10.1039/d4ma00065j

Chapter 7
In-Situ Characterization Techniques for Energy Storage Applications

Yashwanth Venkatraman Naik and D. M. Tejashwini

Abstract Energy storage systems have become crucial in modern society for reducing fossil fuel-related environmental issues and enhancing renewable energy use, with batteries playing a key role by converting electrical energy into chemical energy. The demand for high-energy, high-power rechargeable batteries for portable devices, electric vehicles (EVs), hybrid electric vehicles (HEVs), and large-scale energy storage systems (ESSs) has expanded their applications. Batteries comprise two electrodes, an electrolyte, and an ion-conductive separator, with performance reliant on the electrochemical activity and structural evolution of electrode materials, categorized into insertion-type, alloying-type, and conversion-type based on ion storage behavior. Enhancing battery performance involves techniques like nanostructuring, doping, and surface modification, but future demands require a better understanding of electrochemical ion storage mechanisms. In-situ characterization techniques provide real-time insights into structural and electronic changes in electrode materials, bridging the gap between current and desired battery performance. Key in-situ techniques include X-ray diffraction (XRD), X-ray absorption spectroscopy (XAS), electron microscopy (TEM, SEM, AFM), electrochemical impedance spectroscopy (EIS), current–voltage (I–V) characteristics analysis, charge diffusion studies, X-ray photoelectron spectroscopy (XPS), and neutron diffraction, all of which play vital roles in advancing energy storage technology.

Keywords EVs · In-situ techniques · Energy storage · Characterizations

Y. V. Naik (✉)
Department of Physics, RV Institute of Technology and Management, Bangalore 560076, India
e-mail: naikyashwanthv@gmail.com

D. M. Tejashwini
Department of Studies in Chemistry, Davangere University, Davangere, Karnataka 577007, India

R. Naik et al. (eds.), *Advanced Nanomaterials for Energy Storage Devices*,
Nanostructure Science and Technology, https://doi.org/10.1007/978-3-031-74730-4_7

7.1 Introduction

In modern society, energy storage systems are crucial for reducing environmental issues related to fossil fuels and promoting the adoption of renewable energy sources. Batteries, which convert electrical energy into chemical energy and vice versa, are a prime example. These devices are essential for powering portable electronics like smartphones and laptops. The growing market for high-energy and high-power rechargeable batteries, especially for next-generation portable devices, electric vehicles (EVs), hybrid electric vehicles (HEVs), and large-scale energy storage systems (ESSs), highlights their expanding significance [1].

Batteries consist of two electrodes, an electrolyte, and an ion-conductive separator. The electrodes are electrochemically active components. During charge and discharge cycles, electrons flow from one electrode to the other through an external circuit, while ions move through the electrolyte to maintain electro-neutrality during redox reactions at both electrodes. The performance of a battery is influenced by many factors, with the electrode materials and their reversible structural evolution during cycling being among the most critical. Electrode materials are categorized into insertion-type, alloying-type, and conversion-type based on their ion storage behavior [2].

Enhancing battery performance involves techniques such as nanostructuring, doping, surface modification, and composite design to improve the ionic and electronic conductivity of electrode materials [3]. Despite significant advancements, current technology still falls short of meeting future application demands. A deeper understanding of the electrochemical ion storage mechanisms within electrode materials is essential for bridging this gap. This understanding can lead to more sophisticated designs that optimize the electrochemical performance and durability of batteries.

In-situ characterization techniques have become essential for gaining real-time insights into the structural and electronic changes in electrode materials during battery operation. Unlike ex-situ techniques, which analyze materials outside their working environment and may introduce artifacts, in-situ methods provide a more accurate representation of the materials' behavior under actual operating conditions [4]. Key in-situ techniques include X-ray diffraction (XRD), X-ray absorption spectroscopy (XAS), electron microscopy (TEM, SEM, AFM), electrochemical impedance spectroscopy (EIS), current–voltage (I–V) characteristics analysis, charge diffusion studies, X-ray photoelectron spectroscopy (XPS), and neutron diffraction. These techniques are crucial for advancing energy storage technology by providing detailed insights that guide the optimization and design of high-performance battery materials [5].

7.2 Significance of In-Situ Techniques in Energy Storage

In-situ characterization techniques are vital for advancing the science and technology of energy storage. They offer unparalleled insights into the real-time behavior of electrode materials, enabling performance optimization and the development of next-generation energy storage systems. The significance of these techniques lies in their ability to provide a deeper understanding of the complex processes that govern the functionality and reliability of energy storage devices, thereby driving innovation and progress in this critical field.

7.2.1 Understanding Structural Evolution

During charge and discharge cycles, electrode materials undergo significant structural changes that can affect their performance and lifespan [6]. In-situ techniques, such as X-ray diffraction (XRD) and electron microscopy, allow researchers to track these structural transformations, providing insights into phase changes, lattice distortions, and the development of microstructural features that impact the material's electrochemical behavior.

7.2.2 Investigating Surface and Interface Phenomena

The interfaces between electrodes and electrolytes play a pivotal role in the performance of energy storage devices. In-situ techniques, such as X-ray photoelectron spectroscopy (XPS) and electron microscopy, can probe these interfaces to reveal the formation of solid–electrolyte interphases (SEI), surface reactions, and degradation mechanisms [7]. Understanding these phenomena is essential for improving interface stability and overall device performance.

7.2.3 Real-Time Monitoring of Electrochemical Reactions

In-situ techniques enable the observation of electrochemical reactions as they occur within the battery, providing valuable information on reaction kinetics, ion transport mechanisms, and the formation and dissolution of phases. This real-time monitoring is critical for identifying the factors that influence the efficiency and stability of energy storage systems. In-situ characterization techniques play a crucial role in advancing energy storage technology by providing real-time insights into the mechanisms of ion storage, charge transfer, and material degradation in electrode materials. These insights guide the design and optimization of electrode materials through strategies such as doping, nanostructuring, and surface modification, enhancing their performance and longevity [8]. The empirical data obtained from in-situ studies validate and refine theoretical models, creating a synergy that leads to a comprehensive understanding of material behavior and the development of accurate, predictive models for energy storage applications. Moreover, in-situ techniques address critical challenges in energy storage, such as improving cycle life, increasing energy density, and ensuring safety, by offering detailed information on failure mechanisms and degradation pathways. This knowledge informs the creation of more robust and reliable energy storage solutions, driving innovation and progress in the field [9].

7.3 In-Situ Characterization Techniques for Energy Storage Applications

An electrode material with high-energy density and long-term stability is essential for next-generation applications such as electric vehicles and large-scale energy storage systems. Since the introduction of lithium and sodium-ion batteries, numerous efforts have been dedicated to enhancing the electrochemical properties of electrode materials. Even with some successes, a fundamental understanding of the nature of electrode materials in their working condition still needs improvement [10]. Recent advancements in in-situ X-ray techniques have allowed us to better understand real-time structural changes in electrode materials, clearly explaining their corresponding changes in electrochemical properties [11].

7.3.1 In-Situ XRD Analysis for Active Electrode Material

In-situ X-ray diffraction (XRD) analysis is a vital tool (Fig. 7.1) for understanding and mitigating battery degradation mechanisms in active electrode materials. Degradation during cycling, caused by factors like particle cracking, lithium retention, electrolyte degradation, lithium plating, and dendrite formation, can severely limit the performance and lifespan of battery materials. In-situ XRD allows researchers to observe real-time changes in the crystal structure of electrode materials as the battery charges and discharges. This technique provides crucial insights into how the lattice structure expands and contracts during these cycles, leading to mechanical stresses and eventual particle cracking [12]. For instance, during the charging cycle of a battery, positive lithium ions leaving the cathode cause the crystal lattice to expand, while high-voltage cycling can lead to phase transformations and abrupt lattice expansions, as seen in materials like NCM811 [13, 14]. Similarly, in anodes such as graphite, the lattice structure undergoes changes from graphite to LiC12 and LiC6 during charging, resulting in lithium-ion retention and reduced discharge capacity over time [15].

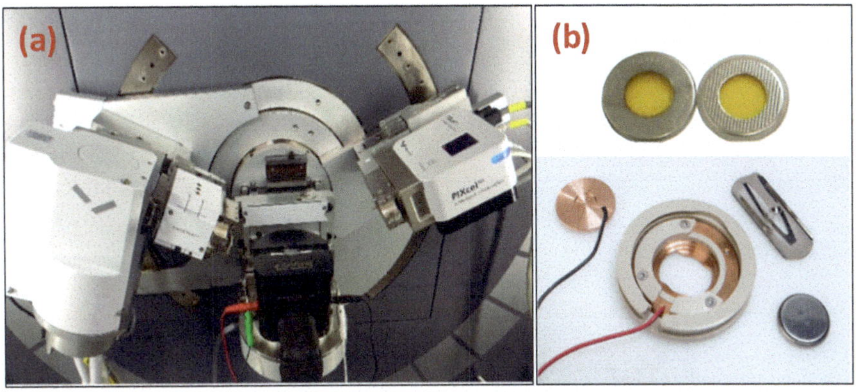

Fig. 7.1 **a** Illustrates an electrochemical cell mounted on the Empyrean XRD platform, which can be heated or cooled to study cycling durability at various temperatures. **b** Depicts a coin cell holder equipped with electrical connections, designed for mounting coin cells that feature at least one face with an X-ray transparent window

The Empyrean XRD platform enhances the capability of in-situ XRD by enabling operando cycling of various types of battery cells, including coin, electrochemical, pouch, and prismatic cells. This platform allows a detailed examination of the electrode materials' stability throughout cycling. For example, in operando cycling of NCM333 cathodes and graphite anodes can be monitored, revealing voltage variations and corresponding changes in lattice parameters [16]. The Empyrean XRD supports studies on coin cells with X-ray transparent windows and provides specialized holders for charge–discharge cycling. Additionally, it supports electrochemical cells with X-ray transparent windows, such as beryllium or glassy carbon, and even enables the analysis of multilayer pouch cells using high-intensity Ag radiation. This comprehensive capability allows battery manufacturers and researchers to systematically investigate and develop new battery materials with enhanced performance and longevity by providing real time, atomic-scale insights into structural changes during operation [17].

7.3.2 In-Situ XAS Analysis for Active Electrode Material

In-situ X-ray absorption spectroscopy (XAS) has emerged as a crucial method for probing the dynamic structures of electrode materials, especially those used in electrocatalysis and battery applications. XAS provides detailed information about both the electronic and structural properties of materials by examining the absorption cross-section versus incident energy. The spectrum is divided into two main regions: X-ray Absorption Near Edge Structure (XANES) and Extended X-ray Absorption Fine Structure (EXAFS) [18]. XANES, which occurs around the absorption edge, reveals information about electronic transitions from core levels to unoccupied states,

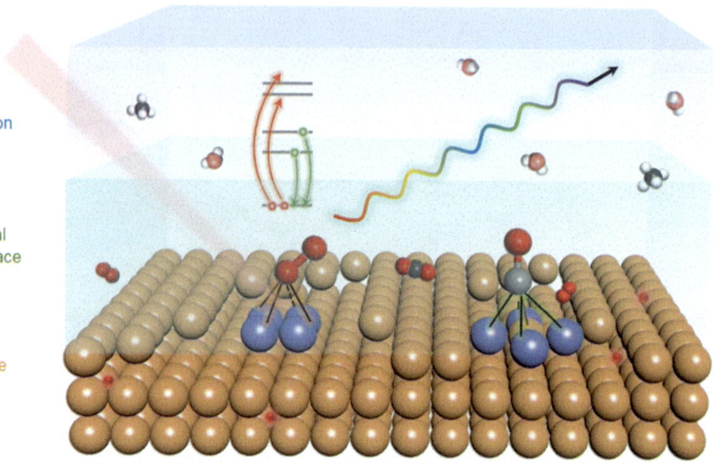

Fig. 7.2 Schematic representation of an electrochemical solid–liquid interface at the molecular level

providing insights into the electronic structure and oxidation states of the elements involved. This region is particularly sensitive to the hybridized states caused by adsorbed reactants and intermediates during electrochemical reactions, making it invaluable for understanding the behavior of electrode materials under operational conditions [19].

EXAFS, located beyond the absorption edge, is characterized by oscillations in the absorption signal due to interference from backscattered photo-electrons from neighboring atoms (Fig. 7.2). This region provides detailed information about the local coordination environment of the absorbing atoms, including the types of neighboring elements, their coordination numbers, and interatomic distances. By analyzing these features, researchers can gain insights into the structural changes and local order around specific atoms during battery cycling. This capability is essential for understanding how structural dynamics at the atomic level influence the performance and stability of electrode materials. In-situ XAS thus plays a pivotal role in the development and optimization of high-performance battery materials, enabling the real-time observation of changes during operation and helping to identify strategies to mitigate degradation and improve overall battery performance [20].

7.3.3 In-Situ Electron Microscopy Analysis for Active Electrode material

To improve battery performance and optimize battery design, it is vital to understand the microstructural evolution and reaction kinetics during cell operation. In recent years, various in-situ/operando techniques have been applied to the reaction mechanism studies of batteries, such as in-situ X-ray diffraction, transmission electron microscopy (TEM), Raman spectroscopy, and nuclear magnetic resonance spectroscopy. Among these techniques, in-situ TEM has been recognized as a powerful tool to track dynamic phase transformation as well as morphological and structural evolution of electrode materials on the atomic scale in real time. This technique can provide significant insights for understanding the fundamental operation principles of battery devices [21]. In-situ TEM for battery investigations includes two configurations: sealed-cell and open-cell. The sealed-cell is used for investigating liquid electrolyte-based rechargeable batteries, while the open-cell can operate in solid or ionic liquid electrolytes. Each configuration has its own advantages and limitations. The structural evolution, phase transformation, reaction kinetics, solid–electrolyte interface (SEI), and interfacial behavior of sodium-ion batteries (SIBs) are directly correlated with electrochemical properties. In-situ TEM is an effective technique to investigate these behaviors in real time, especially the reaction kinetics, ion diffusion, and SEI growth. Different anode materials exhibit various Na-ion storage behaviors. For instance, hard carbon undergoes intercalation with negligible volume expansion, while alloy-type anode materials such as Sn, Sb, and phosphorus exhibit significant volumetric variation [22]. The morphological changes and volume expansion during cycling were extensively investigated using in-situ TEM to correlate with electrochemical properties. The in-situ observation of various materials in sodium-ion batteries provides critical insights into their fundamental reaction mechanisms, guiding the design and development of high-performance electrode materials. As in-situ TEM capabilities advance, it is expected that these techniques will significantly impact the future design and optimization of battery materials [17].

7.3.4 In-Situ Electrochemical Impedance Spectroscopy Analysis for Active Electrode Material

Electrochemical impedance spectroscopy (EIS) is crucial for understanding battery mechanisms but often faces challenges with low-frequency data complexity. To address this, we conducted a detailed near-in-situ EIS analysis on lithium iron phosphate (LFP) cells, which exhibit stable charge/discharge platforms and distinct reaction voltage stages. By applying the macrohomogeneous porous electrode theory, we accurately analyzed low-frequency impedance features, establishing equivalent circuit models for each stage. This approach yielded ionic and electronic resistivities that matched actual cell performance, enhancing our understanding of LFP cells. Our findings provide a foundation for advancing EIS applications in battery systems, optimizing design, and improving battery management [23].

7.3.5 In-Situ Current–voltage (I–V) Characteristics Analysis for Active Electrode Material

The solid–electrolyte interphase (SEI) layer, forming between the active electrode and electrolyte during battery operation, is critical to rechargeable battery performance due to its role in preventing parasitic reactions while facilitating ionic conduction for Faradaic reactions. Despite its importance, the electrical properties of SEI layers have been poorly understood due to the lack of direct characterization methods. Utilizing in-situ bias transmission electron microscopy, we directly measured the electrical properties of SEI layers formed on copper (Cu) and lithium (Li) substrates. Our findings revealed that SEI behaves as a voltage-dependent differential conductor rather than a typical electrical insulator, challenging long-held assumptions [24]. The thickness and topography of SEI, crucial for electrochemical properties, are influenced by the differential conductance rate; higher rates lead to thicker, more complex SEI structures, which reduce Coulombic efficiency and cycling stability. We also discovered that SEI with high inorganic content has high electron resistance, whereas SEI rich in organic species shows higher electron leakage. These insights were corroborated by reactive force field and hybrid ab initio molecular dynamics simulations. This study resolves the long-standing mystery of SEI's electrical behavior during electrochemical operations, identifying differential conductance as a key factor influencing SEI and Li deposit characteristics, thereby informing the design of SEI for improved battery performance [25].

7.3.6 In-Situ Charge Diffusion Studies Analysis for Active Electrode Material

The diffusion process and mechanism are crucial to the lithium storage performance of electrode materials, significantly impacting the electrochemical performance of lithium-ion batteries. This paper presents an in-situ experimental method based on phase evolution to observe the diffusion process. An optical acquisition system with an electrochemical battery at its core was constructed, utilizing the electrochromic characteristics of graphite electrodes to display phase evolution information in real time [26]. Diffusion process observation experiments at various charge/discharge rates (C-rates) revealed a rate-dependent relationship among potential, phase transition, and lithium content. Results indicate that high C-rates accelerate the phase evolution process, lower the formation potential of new phases, shorten phase transition duration, and lead to a multiphase coexistence state in the electrode. The acceleration effect induced by high C-rates is staged, with significant acceleration in the early and middle stages, and slower acceleration later on. This is due to the competition between diffusion-driving energy and the energy barrier. Therefore, an optimization strategy balancing battery capacity and lithiation rate is proposed to enhance performance [27].

7.3.7 *In-Situ X-ray Photoelectron Spectroscopy Analysis for Active Electrode Material*

Understanding charge transfer processes at electrochemical interfaces is crucial for designing advanced energy storage materials. While X-ray photoelectron spectroscopy (XPS) is pivotal in analyzing electronic structures of interfaces like electrode–electrolyte interphases, ex-situ studies based on postmortem analysis often miss intermediate and transient species essential for predicting charge transfer processes [28]. The scarcity of in-situ/operando XPS studies on buried interfaces in energy storage systems stems from technical challenges, such as maintaining high vacuum conditions. However, significant advancements have been made in the past decade to overcome these barriers, allowing investigation of solid/solid and solid/liquid interfaces. This review highlights recent progress and new experimental designs in applying in-situ and operando XPS to characterize interfacial processes and emergent properties [29]. These advancements can inform the design strategy for advanced batteries. The review also discusses the remaining challenges and future research directions, outlining potential paths forward in this field.

7.3.8 *In-Situ Neutron Diffraction for Active Electrode Material*

Electrochemistry and diffraction techniques complement each other exceptionally well for investigating solids and their surfaces, with electrochemical experiments providing thermodynamic and kinetic information and diffraction offering insights into atomic structures [30]. In-situ neutron diffraction allows the simultaneous acquisition of these types of data during phase transitions at ambient temperatures, enabling the tracking of rapid changes and the discovery of otherwise inaccessible metastable and transitional phases and structures. This method also yields valuable information on electrode surfaces, making it particularly useful in battery research. However, its application extends beyond battery materials [31]. This short review provides a practical guide for applying in-situ neutron diffraction in electrochemical experiments, discussing the methodology, available setups, and presenting recent examples to illustrate its utility.

7.4 Conclusion

In-situ characterization techniques are essential tools for the advancement of energy storage applications. These methods provide real-time insights into the structural and electronic changes occurring in electrode materials during operation, thus bridging the gap between current performance and future technological demands. Techniques such as X-ray diffraction (XRD), X-ray absorption spectroscopy (XAS), electron microscopy (TEM, SEM, AFM), electrochemical impedance spectroscopy (EIS), current–voltage (*I*–*V*) characteristics analysis, charge diffusion studies, and neutron diffraction enable a detailed understanding of ion storage mechanisms, charge transfer processes, and material degradation pathways. This comprehensive understanding is crucial for the design and optimization of electrode materials, leading to enhanced battery performance, longer cycle life, increased energy density, and improved safety. By addressing critical challenges and validating theoretical models with empirical data, in-situ characterization techniques promote the development of more robust and reliable energy storage solutions. Ultimately, these advancements contribute to the progression of renewable energy technologies and the reduction of environmental impacts associated with fossil fuels.

References

1. Ahmad F, Khalid M, Panigrahi BK (2021) Development in energy storage system for electric transportation: a comprehensive review. J Energy Storage 43:103153. https://doi.org/10.1016/j.est.2021.103153
2. Prajapati AK, Bhatnagar A (2023) A review on anode materials for lithium/sodium-ion batteries. J Energy Chem 83:509–540. https://doi.org/10.1016/j.jechem.2023.04.043
3. Tiwari VK, Singh RK (2023) Nanostructured coating strategies of cathode for improved sodium ion battery performance. Chem Eng J 471:144592. https://doi.org/10.1016/j.cej.2023.144592
4. Sangid MD (2020) Coupling in situ experiments and modeling—opportunities for data fusion, machine learning, and discovery of emergent behavior. Curr Opin Solid State Mater Sci 24(1):100797. https://doi.org/10.1016/j.cossms.2019.100797
5. Sun J, Wu M, Jiang H, Fan X, Zhao T (2021) Advances in the design and fabrication of high-performance flow battery electrodes for renewable energy storage. Adv Appl Energy 2:100016. https://doi.org/10.1016/j.adapen.2021.100016
6. Li R, Li W, Singh A, Ren D, Hou Z, Ouyang M (2022) Effect of external pressure and internal stress on battery performance and lifespan. Energy Storage Mater 52:395–429. https://doi.org/10.1016/j.ensm.2022.07.034
7. Wu B, Wang B, Petit T (2021) Soft X-ray spectroscopy of light elements in energy storage materials. Energy Storage Mater 40:72–95. https://doi.org/10.1016/j.ensm.2021.04.045
8. Shah SS, Niaz F, Ehsan MA, Das HT, Younas M, Khan AS, Rahman HU, Nayem SA, Oyama M, Aziz MA (2024) Advanced strategies in electrode engineering and nanomaterial modifications for supercapacitor performance enhancement: a comprehensive review. J Energy Storage 79:110152. https://doi.org/10.1016/j.est.2023.110152
9. McIlwaine N, Foley AM, Morrow DJ, Al Kez D, Zhang C, Lu X, Best RJ (2021) A state-of-the-art techno-economic review of distributed and embedded energy storage for energy systems. Energy 229:120461. https://doi.org/10.1016/j.energy.2021.120461
10. Poizot P, Gaubicher J, Renault S, Dubois L, Liang Y, Yao Y (2020) Opportunities and challenges for organic electrodes in electrochemical energy storage. Chem Rev 120(14):6490–6557. https://doi.org/10.1021/acs.chemrev.9b00482
11. Chang CJ, Zhu Y, Wang J, Chen HC, Tung CW, Chu YC, Chen HM (2020) In situ X-ray diffraction and X-ray absorption spectroscopy of electrocatalysts for energy conversion reactions. J Mater Chem A 8(37):19079–19112. https://doi.org/10.1039/D0TA06656G
12. de Vasconcelos LS, Xu R, Xu Z, Zhang J, Sharma N, Shah SR, Han J, He X, Wu X, Sun H, Hu S, Perrin M, Wang X, Liu Y, Lin F, Zhao K (2022) Chemomechanics of rechargeable batteries:

status, theories, and perspectives. Chem Rev 122(15):13043–13107. https://doi.org/10.1021/acs.chemrev.2c00002

13. Wu Z, Zhang C, Yuan F, Lyu M, Yang P, Zhang L, Zhou M, Wang L, Zhang S, Wang L (2024) Ni-rich cathode materials for stable high-energy lithium-ion batteries. Nano Energy 109620. https://doi.org/10.1016/j.nanoen.2024.109620
14. Liu J, Zou Z, Zhang S, Zhang H (2021) Structure, modification, and commercialization of high nickel ternary material ($LiNi_{0.8}Co_{0.1}Mn_{0.1}O_2$ and $LiNi_{0.8}Co_{0.15}Al_{0.05}O_2$) for lithium ion batteries. J Solid State Electrochem 25:387–410. https://doi.org/10.1007/s10008-020-04818-5
15. Shin H, Lee YK, Lu W (2022) Structural degradation of graphite anode induced by dissolved manganese ions in lithium-ion batteries. J Power Sources 528:231223. https://doi.org/10.1016/j.jpowsour.2022.231223
16. Chakraborty A, Kunnikuruvan S, Kumar S, Markovsky B, Aurbach D, Dixit M, Major DT (2020) Layered cathode materials for lithium-ion batteries: review of computational studies on $LiNi_{1-x-y}Co_xMn_yO_2$ and $LiNi_{1-x-y}Co_xAl_yO_2$. Chem Mater 32(3):915–952. https://doi.org/10.1021/acs.chemmater.9b04066
17. Yousaf M, Naseer U, Li Y, Ali Z, Mahmood N, Wang L, Gao P, Guo S (2021) A mechanistic study of electrode materials for rechargeable batteries beyond lithium ions by in situ transmission electron microscopy. Energy Environ Sci 14(5):2670–2707. https://doi.org/10.1039/D0EE03295F
18. Iglesias-Juez A, Chiarello GL, Patience GS, Guerrero-Pérez MO (2022) Experimental methods in chemical engineering: X-ray absorption spectroscopy—XAS, XANES, EXAFS. Can J Chem Eng 100(1):3–22. https://doi.org/10.1002/cjce.24291
19. Chen C, Jin H, Wang P, Sun X, Jaroniec M, Zheng Y, Qiao SZ (2024) Local reaction environment in electrocatalysis. Chem Soc Rev. https://doi.org/10.1039/D3CS00669G
20. Wu Z, Kong Pang W, Chen L, Johannessen B, Guo Z (2021) In situ synchrotron X-ray absorption spectroscopy studies of anode materials for rechargeable batteries. Batteries Supercaps 4(10):1547–1566. https://doi.org/10.1002/batt.202100006
21. Yu L, Hudak BM, Ullah A, Thomas MP, Porter CC, Thisera A, Pham RH, De Alwis Goonatilleke M, Guiton BS (2020) Unveiling the microscopic origins of phase transformations: an in situ TEM perspective. Chem Mater 32(2):639–650. https://doi.org/10.1021/acs.chemmater.9b03360
22. Wang G, Yu M, Feng X (2021) Carbon materials for ion-intercalation involved rechargeable battery technologies. Chem Soc Rev 50(4):2388–2443. https://doi.org/10.1039/D0CS00187B
23. Buchicchio E, De Angelis A, Santoni F, Carbone P, Bianconi F, Smeraldi F (2023) Battery SOC estimation from EIS data based on machine learning and equivalent circuit model. Energy 283:128461. https://doi.org/10.1016/j.energy.2023.128461
24. Xu Y, Jia H, Gao P, Galvez-Aranda DE, Beltran SP, Cao X, Le PM, Liu J, Engelhard MH, Li S, Wang C (2023). Direct in situ measurements of electrical properties of solid–electrolyte interphase on lithium metal anodes. Nat Energy 8(12):1345–1354. https://doi.org/10.1038/s41560-023-01361-1
25. Cao W, Li Q, Yu X, Li H (2022) Controlling Li deposition below the interface. EScience 2(1):47–78. https://doi.org/10.1016/j.esci.2022.02.002
26. Shi B, Han B, Xie H, Kang Y, Zhang Q (2021) C-rate related diffusion process of the graphite electrode by in situ experiment and analysis. Electrochim Acta 378:138151. https://doi.org/10.1016/j.electacta.2021.138151
27. Deng W, Phung J, Li G, Wang X (2021) Realizing high-performance lithium-sulfur batteries via rational design and engineering strategies. Nano Energy 82:105761. https://doi.org/10.1016/j.nanoen.2021.105761
28. Rehman S, Pope M, Tao S, McCalla E (2022) Evaluating the effectiveness of in situ characterization techniques in overcoming mechanistic limitations in lithium–sulfur batteries. Energy Environ Sci 15(4):1423–1460. https://doi.org/10.1039/D1EE03396D
29. Thapa S, Paudel R, Blanchet MD, Gemperline PT, Comes RB (2021) Probing surfaces and interfaces in complex oxide films via in situ X-ray photoelectron spectroscopy. J Mater Res 36:26–51. https://doi.org/10.1557/s43578-020-00070-9

30. Magnussen OM, Drnec J, Qiu C, Martens I, Huang JJ, Chattot R, Singer A (2024) In situ and Operando X-ray scattering methods in electrochemistry and electrocatalysis. Chem Rev 124(3):629–721. https://doi.org/10.1021/acs.chemrev.3c00331
31. Li J, Fleetwood J, Hawley WB, Kays W (2021) From materials to cell: state-of-the-art and prospective technologies for lithium-ion battery electrode processing. Chem Rev 122(1):903–956. https://doi.org/10.1021/acs.chemrev.1c00565

Chapter 8
Case Studies: Nanomaterials in Specific Energy Storage Devices

B. M. Santhosh, H. P. Shivarudrappa, B. C. Vasantha Kumar, P. N. Prashantha Kumar, Soniya D. Naik, Prasannakumar Jammapra Kallappa, and T. Divya

Abstract The chapter explores the revolutionary role of nanotechnology in enhancing energy storage solutions, focusing on the advancements in lithium-ion batteries (LIBs), supercapacitors, sodium-ion batteries (SIBs), and zinc-air batteries. As the world faces the dual challenges of energy sustainability and climate change, the transition to renewable energy sources has become crucial. The intermittent nature of renewable energy sources like wind and solar necessitates advanced energy storage technologies to balance supply and demand effectively. Silicon nanowires for LIB anodes, with a theoretical charge capacity of 4200 mAh/g, address volume change issues during lithiation and delithiation, achieving specific capacities of about 3100 mAh/g after 10 cycles. Graphene-based supercapacitors leverage graphene's exceptional electrical conductivity and surface area, showing high specific capacitance and excellent cycle stability. Titanium dioxide nanotubes for SIB anodes enhance performance by accommodating sodium's larger ionic radius, achieving reversible capacities of approximately 300 mAh/g with stable cycling. Cobalt oxide nanoparticles have their place in the development of zinc-air batteries as a result

B. M. Santhosh · H. P. Shivarudrappa (✉)
Department of Chemistry, G.M. University, P.B. Road, Davanagere, Karnataka 577006, India
e-mail: dr.shivahp@gmail.com

B. C. Vasantha Kumar · P. N. Prashantha Kumar
Department of Chemistry, D.R.M Science College, Davanagere, Karnataka 577004, India

S. D. Naik
Biorefining Conversions and Fermentation Laboratory, Agriculture/Forestry Centre, University of Alberta, Edmonton, Canada

P. Jammapra Kallappa
Department of Chemistry, Bapuji Institute of Engineering and Technology, Davanagere, Karnataka 577004, India

T. Divya
Department of Chemistry, A.V. Kamalamma College for Women, Davanagere, Karnataka 577002, India

B. M. Santhosh
Visvesvaraya Technological University, Belagavi, India

R. Naik et al. (eds.), *Advanced Nanomaterials for Energy Storage Devices*, Nanostructure Science and Technology, https://doi.org/10.1007/978-3-031-74730-4_8

of the successful transfer of reducing agents to oxygen molecules and the subsequent anticipated oxidation of these oxygen molecules which, in turn, considerably increases energy efficiency and cycle stability. The progress of these nanotechnologies in energy storage is one of the basic factors to be taken care of in order to enable the sustainability of the future, as the overcoming of the challenges of scalability and cost effectiveness is necessary for their widespread adaptation.

Keywords Nanomaterials · Energy storage devices

8.1 Introduction

As the world grapples with the dual challenges of energy sustainability and climate change, the transition to renewable energy sources has become imperative. A sustainable energy supply necessitates a significant shift from fossil fuels to renewable sources. By 2050, it is anticipated that at least 60% of Germany's gross final energy consumption will be sourced from renewables, assuming current supply security demands remain constant [1]. However, this transition is not without its challenges, particularly the natural fluctuations in power output from renewable sources like wind and solar power. To ensure a consistent supply of electrical energy, various flexibility options must be employed to balance supply and demand. These include grid expansion, load management, the use of highly flexible conventional power plants, and the deployment of advanced energy storage methods.

Energy storage technologies play a pivotal role in this balancing act. These technologies encompass a wide range of methods classified by the type of energy stored, including thermal energy (heat accumulators), chemical energy (accumulators, batteries, redox flow batteries, hydrogen, methane), mechanical energy (flywheels, springs, pumped-storage power plants), and electrical energy (capacitors and superconducting magnetic energy storage). The storage of electricity is inherently complex, as it requires conversion into other energy forms, which incurs losses during both storage and reconversion phases. Despite these challenges, innovations in nanotechnology are poised to revolutionize energy conversion, storage, and transmission.

Nanotechnology offers ground-breaking solutions for enhancing energy storage and conversion efficiency [2]. The application of nanomaterials, with their unique properties and versatile applications, has opened new avenues in energy technology. For instance, in the realm of rechargeable lithium batteries, silicon nanowires have emerged as a promising material for anodes. Silicon boasts a low discharge potential and the highest known theoretical charge capacity (4200 mAh/g), significantly surpassing the capacities of traditional graphite anodes and various other materials. However, silicon's practical application has been limited due to its volume changes (up to 400%) during lithiation and delithiation, which cause pulverization and capacity fading. Silicon nanowire battery electrodes address these issues by accommodating large strains without pulverization, maintaining good electronic contact, and providing efficient charge transport [3, 4].

Nanoparticles have made notable contributions to solar energy conversion and supercapacitors. Carbon-based nanoparticles, such as carbon nanotubes, activated carbon, and graphene, are highly impressive, especially owing to their significant surface area, porosity tuning, and thus, making them have strong energy density (W). These properties make them ideal for rapid energy storage and discharge applications. A major area that nanotechnology has contributed to is sustainable energy through solar energy production. Nanoparticles are able to increase photovoltaic (PV) efficiency by light trapping [5], which is one of a kind plasmonic effect and even results in high-performance solar cells by greater and more efficient carrier diffusion. The quantum dot semiconductor particles and, particularly, the (CdSe) ones, play a leading role in the development of new energy collection technologies by promoting carrier dissociation processes [6].

Catalysis is just one example of where nanoparticles do an amazing job. They depend on the "machine" size of the nanoparticles, which dislocates electronic properties, and the availability of a large space of their surfaces, which makes extraordinarily adjacent the reactant molecules, resulting in elevated catalysis that is truly remarkable for processes such as hydrogen production, oxygen reduction, and carbon dioxide conversion. These strides made in catalysis are not only significant for sustainable energy technologies but they also portray the ripple effect that will occur in the energy landscape further exposing the potential of nanoparticles that would cause the whole energy filtrate to crack [7, 8].

Nevertheless, their potential, nanoparticles encounter some hurdles that need to be solved for practicing implementation. Scale, stability, and cost-effectiveness will be the biggest challenges operating under the old system of trying to solve the issues with nanoscience techniques. Nevertheless, nanoparticles never cease to show new perspectives into energy-related fields. After being dispersed on the surface, nanoparticles will promote the electrode electrochemical surface growth, which in turn will deliver higher battery capacity and faster charge/discharge rates. In, photovoltaic cells, use of nanoparticles like gallium arsenide and indium phosphide for light absorption and conversion efficiency, enhancing energy exchange shows that dots could really make a difference in green technology applications. Quantum dots in LED lighting prove to be a great deal, as they will enable the lights to be controlled most suitably and thus, take up less modern power consumption in the end. Nanoparticles are also applicable to optimizing chemical processes using high-performance catalysts in a variety of industrial applications.

Energy storage is the backbone of the shift towards renewables [9] energy sources and the global mobile electronics phenomenon. The penetration of lithium-ion, supercapacitor technology, and the shot of power given by the discovery of systems like zinc-air and sodium-ion to the market is unparalleled. The incorporation of nanomaterials into these energy storage devices has really changed the performance game, providing superior energy density, high charge/discharge rates, and long cycle life. The section discusses various examples by constructing the effect of thermal measurement of nanomaterials on the electrodes, working out the revolution of nanotechnology both in text and illustrations that give a unique and tangible example of achievements are made [10].

8.2 Silicon Nanowires for High-Capacity Anodes in Lithium-Ion Batteries

The flexibility of LIBs has led to the development of energy storage devices and hence the powering of a wide range of devices including xEV, which are expected to grow toward lead-acid batteries and other ultra-light supercapacitors. The conventional graphite anode in LIBs has a storage capacity limit of about 372 mAh/g, which is quite good. The introduction of silicon on the other hand, with up to 4200 mAh/g, is a breakthrough compared to that of graphite. Both however, the promotion of silicon anodes was limited as a result of SOC changes during lithiation (around 300%) and delithiation. Systems changes often trigger mechanical failure and capacity fade after a short time, thus presenting major technical problems in their practical application.

The design of these nanostructures has allowed to accommodate the volume expansion and contraction needed during the charge–discharge cycle, thus, the improvement of the battery life and performance. Examples of the structure of silicon nanowires as well as the lithiation process and the way the nanowires can be used to solve the problem of mechanical stresses that bulk silicon anodes cannot handle are given below [11].

The sensational breakthrough in the LIB performance metrics stands with the debut of silicon nanowire anodes. A study carried out by Chan et al. (2008) was the turning point in the use of silicon nanowires as anodes. In this study, it was noticed that the new silicon anode can achieve a specific capacity of 3 100 mAh/g after 10 cycles, while the traditional graphite anodes could only 370 mAh/g. This is a very impressive difference. Again, the light of this result is the fact that silicon anodes are incomparably potential for practical applications in the field of energy storage systems.

8.2.1 Detailed Analysis

1. **Electrochemical Performance**

High Surface Area: The silicon nanowire concept is quite reliable, which allows for a big surface area that in turn improves the transportation of lithium ion and the rate of the charge/discharge process. The distance the lithium ions have to travel through is the main issue in this technology. This company promises this feature. Although electron transfer is the fastest way to improve electrochemical reactions, the specific capacity of battery components is also related to the overall battery cycle performance, which in turn is correlated with the capacity of the charging source.

2. **Mechanical Stability**

Radial Expansion Accommodation: This is achieved due to the fact that the silicon nanowires are able to grow radially. The one-dimensional structure of silicon nanowires allows them to expand radially. This ability to accommodate volumetric changes without fracturing is crucial, as it prevents the mechanical degradation typically observed in bulk silicon anodes. This structural resilience is a key factor in the longevity and reliability of silicon nanowire-based batteries.

Durability: The elasticity of nanowires due to the repeated elongation and shrinkage cycles they take goes a long way in keeping

3. **Cycle Life**

Prolonged Durability: The continual reliability maintained by silicon nanowire anodes needs to be pretty significant for a long time to allow for a long cycle life. In the end, the silicon nanowire anodes will be stronger which will enable them to be used in commercial applications.

Capacity Retention: Though it is still in the trial phase, silicon nanowire anodes seem to hold on to their capacity well over a number of cycles. Thus, they are best suited for applications that require equipment to be used consistently over time and to the maximum efficiency of each cycle they can offer.

8.2.2 Challenges and Future Directions

However, the potential of technology to the journey stage or more will be utilized if, these silicon nanowire anodes face several issues that must be tackled to make them commercially something.

8.2.3 High Production Costs

Manufacturing Complexity: The making of silicon nanowires is not merely costly but also a demanding task that cannot be carried out easily in large quantities. The heart of the matter lies in creating methods which are resource-effective, have minimal costs and at the same time are capable of enlarging the number of silicon nanowire anodes for the market in the future.

8.2.4 Material Costs

The raw materials and emitting silica are the ones that make people pay a lot and the processes of creating silica nanowires also add up to the technology charge.

8.2.5 Integration of Commercial Batteries

Compatibility with Existing Systems: Incorporating the ones that are in the already manufactured batteries along with silicon nanowire anodes is something that the company has to manage without impairing the performance. The essential thing is that this complex can be accomplished only through the combination of a capacity of the right technological engineering that is not detrimental to current production and a battery architecture able to be equipped with such devices.

Performance Optimization: Providing that the increased capacity of silicon nanowire anodes serves well in practice, the problem demands further research and development.

Further research is dedicated to attaining this goal by the development of more effective technique of synthesis and the optimization of integrating the silicon nanowire into commercial battery systems. Nanotechnological, supermaterial, etc., have led to the opening of untapped sciences, such as the development of silicon, gold, ZnO, for example, segue synthesis, the nanomaterials, advanced materials, and nanotechnology, among others, are really blooming the things that synthetic organic chemistry came about for. Batteries can now have silicon nanowire anodes with high capacity, long life that can be produced by improving new technological materials. However, the innovation in this industry will cause silicon nanowires to support next-generation energy storage systems, which will be the backbone of the future power systems that will satisfy the need for ever-growing technological advancement and will be in line with our vision of a greener and more sustainable energy future.

8.3 Supercapacitors: Graphene-Based Innovations

Supercapacitors, characterized by their extremely high power density and long cycle life, are the best candidates for a wide range of high-performance energy storage applications. The use of nanomaterials, in particular graphene, has greatly added to the functionality of supercapacitors. Graphene, a single layer of carbon atoms arranged in a hexagonal lattice, stands out due to its enormous electrical conductivity, mechanical strength, and disproportionately large surface area which could be as high as 2630 m^2/g. This chapter dives into the incredible influence of graphene on supercapacitor technology and refers to a classic study by Stoller et al., which showed the much better performance of rGO in supercapacitors than other materials [12].

8.3.1 Graphene's Unique Properties

Graphene's unique characteristics are the reason for which it gets the full attention as a potential supercapacitor enhancer. The high surface area of the material indicates the existence of a large number of active sites for ion adsorption and desorption, which significantly enhances the capacitance. In addition, the good electrical conductivity of the graphene material would ensure efficient charge transport within the electrode, which is very important to high power density. Besides, graphene's mechanical strength and flexibility also enable the development of durable and flexible supercapacitor designs, which would expand their application potential [13].

- **Case Study: Reduced Graphene Oxide Supercapacitors**

In the study proposed the application of rGO sheets in supercapacitors, which revealed that they are much more efficient than the conventional activated carbon electrodes. The rGO-based supercapacitors illustrated a high specific capacitance and rate capability, which was the 'strong case' that was built for graphene energy storage devices. The following parts contain a detailed description and importance of the study.

(1) **High Surface Area**

The great surface area of graphene is the most significant factor that increases the efficiency of supercapacitors. This large surface area gives the electrode a lot of active sites for the electrochemical processes that occur in recharge and discharge cycles. The study by Stoller et al. pointed out that the rGO sheets have a big surface area which is essential for ion adsorption and ion desorption to be efficient. This is a must in gaining a higher capacitance since it makes possible the accumulation of a larger amount of charge within the electrode material.

(2) **Electrical Conductivity**

Graphene's unique electrical conductivity is another advantage for supercapacitors. The charge transport in the electrode material that is highly efficient is central to high power density, as it enables quick charge and discharge cycles. The rGO-based supercapacitors, the development of which was unfolded by Stoller et al., demonstrated outstanding electrical conductivity. This in turn was a consequence of the improvement in the "rate capability" of the chargers. Therefore, supercapacitors can be regarded as powerful tools, as they are able to provide large power outputs in a short period of time that can be used in the application of quick bursts of energy.

(3) **Mechanical Flexibility**

The fact that the mechanical resistance and flexibility of graphene are such that it is flexible and can be additionally used for the production of a supercapacitor with a long lifespan was the conclusion. The fact that this feature is mainly used in portable and wearable electronics, which require flexibility and durability, makes it particularly useful. The rGO supercapacitors, which are the compound of graphene oxide, maintained their structural integrity and performance during many cycles and all such had the potential to operate for a long time in many applications.

8.3.2 Performance Metrics

A research by Stoller et al. focused solely on the impressive performance metrics for rGO-based supercapacitors. For example, the specific capacitance for these supercapacitors in aqueous medium was 135 F/g and as for in organic medium it was 99 F/g. This indicates the significant superiority of the new electrodes over traditional activated charcoal-based ones. Additionally, the rGO-based devices were found to have retained over 90% of their capacitance after 10,000 cycles, indicating excellent cycle stability. The elongated cycle life is a primary requirement for practical applications as it guarantees that the supercapacitors can work for a long time without showing any wear.

8.3.3 Challenges and Future Directions

Notwithstanding the positive results, the challenges that need to be worked out regarding graphene-based supercapacitors area considerable. The primary challenge is the large-scale production of high-quality graphene at the minimum cost. The process of making graphene available is costly and complicated. On such a vast scale, the whole cost story only clouds the use of inherently small and slow technologies. The story in this case is really about how to unlock the higher potential of those technologies. One way of addressing the issue is the use of cost-effective and easy-to-produce frames for graphene to have nearly unlimited use of these materials. Microbodies secrets are part of the research in combining graphene with other nanomaterials to produce the electrodes, which can potentially be the next generation of supercapacitors. To improve the efficiency of supercapacitors, graphene can be combined with other substances like metal oxides, conducting polymers, or other carbon-based nanomaterials. Hybrid structures are easy to build from the constituting parts to performance. In other words, they reflect the sum of the best properties of the parts including capacitance, conductivity, and cycle stability. That is the reason for the difference in structure of the compound being and the compound that is not formed.

8.4 Sodium-Ion Batteries (SIBs): Titanium Dioxide Nanotubes for Anode Applications

Numerous research activities are focused on the use of graphene-like anode materials in electrodes for Li/S batteries, and with new nonaqueous electrolyte compounds. The interest in sodium-ion batteries as a low-cost alternative to lithium-ion batteries in energy storage applications has increased in recent years. Na is a widely available chemical element, its cost is low and the resultant sodium batteries are therefore a very attractive and sustainable solution to the world's energy dilemma. Nonetheless, the larger ionic radius of sodium than lithium, for example, causes many problems in what has become a traditional activity in some materials and thus, it is necessary to develop novel materials to solve this problem. Before I begin mentioning the other points I would like to start with the situation that is produced when an element loses an electron while another receives it. This is the case, for example, in chemical reactions. In this process, called oxidation, electrons are transferred from the substance that is being oxidized to the substance that...

8.4.1 The Promise of Titanium Dioxide Nanotubes

Titanium dioxide (TiO_2) nanotubes are one of the most potential materials for SIB anodes. These nanotubes have numerous advantages because of their quirky and biochemical properties. Wu et al.'s study recognized the potential of TiO_2 nanotubes extracted through anodization for use in SIBs highlighting what it is. A high surface area structure is produced through it that is specifically conducive to sodium-ion intercalation, which in turn boosts the battery's performance [14].

8.4.2 Detailed Analysis of TiO_2 Nanotube Performance

The success of TiO_2 nanotubes being used as anodes in SIBs cannot be restricted to a single positive factor; it is the result of the following key factors:

- **High Surface Area**

The high surface area of TiO_2 nanotubes experiences the strong interactions between them and sodium ions. The extra interaction that is developed is very important for the power of the electrochemical anode. The large surface area facilitates more active sites, thus, more sodium ions are adsorbed and desorbed, which in turn means more capacity and better performance.

- **Efficient Ion Diffusion**

The tubular structure of TiO_2 nanotubes enables smoother ion diffusion. This kind of structure results in lower internal resistance within the anode material, so the battery can be charged and discharged at a higher speed. Ion diffusion is a vital process for the battery to maintain high performance and decrease energy losses during cycling.

- **Structural Stability**

Furthermore, besides the electrochemical stability, the mechanical stability of TiO_2 nanotubes is a significant attribute, which promotes their acceptability as anode materials for SIBs. One of the factors that are often harmful to battery performance is the repeated intercalation and deintercalation of sodium ions. The stress caused by these processes can lead to structural damage and the consequent capacity fading in many materials. In contrast, this notch in this phase of the work is the unique structural setup of TiO_2 nanotubes, which cancel out the aforementioned adversity, thereby increasing the cycle life of the battery.

8.4.3 *Performance Metrics*

One of the performance metrics based on Wu et al. (2011) for TiO_2 nanotube anodes in SIBs is promising. The TiO_2 nanotube anodes with excellent efficiency delivered a reversible capacity of around 300 mAh/g, a fact promoted for sodium-ion storage. The other characteristic that caught attention in the study was sticking to the cycling stability principle due to the capability of TiO_2 nanotubes to maintain their physical and electrochemical properties throughout many charge cycles and use cycles. The durability of this is especially vital in the situations of grid storage, where long-term stability is required.

8.4.4 *Challenges and Future Directions*

The fact that the TiO_2 nanotubes are suitable for the SIB anodes is seen to be the best aspect of it but there are a lot of challenges to be faced to see the real application of it:

- **Scalability**

First, the difficulties in the large-scale production of high-quality TiO_2 nanotubes is one of the bottlenecks to high-energy storage devices. The anodization process, a procedure used to produce these nanotubes, must be changed if these nanotubes are supposed to be made on an ultra-large scale to satisfy the application demands of commercial products. The building of a product through which processing is both scalable and cost-effective is vital for the wider utilization of TiO_2 nanotube anode.

- **Cost-Efficient Synthesis**

TiO_2 nanotube cost could be one of the factors that make SIBs a viable alternative to LIBs. Up to now, the synthesization of nanotubes has often been accomplished in some very complex ways that are too expensive for consumer products. Areas of research are seeking economically efficient production methods that do not degrade the quality or performance of the nanotubes.

- **Hybrid Anode Materials**

With the combination of TiO_2 nanotubes and other nanomaterials in the creation of the hybrid anodes, there are high opportunities for the final discoveries in this scientific area. Hybrid materials can be the best of both worlds where a combination of all provided components can lead to a higher performance although keeping low costs. To illustrate, it is possible to synthesize TiO_2 with carbon-based materials or other metal oxides so that the anodes' electrochemical properties and cycle life are significantly enhanced.

8.5 Zinc-Air Batteries: Advancements with Cobalt Oxide Nanoparticles

Zinc-air batteries have made great leaps in the renewable energy storage field largely because of their big energy packages accompanied by relatively low costs, and friendly nature by the use of non-toxic and abundant materials. These batteries gather the readily available oxygen from the atmosphere, thereby converting it to electricity through an electrochemical process. Thus, they can provide green power storage, while increasing the efficiency and durability are crucial for the longevity of the batteries. The OER and the ORR (oxygen reduction reaction) are the reactions that most affect their operation. The OER is a positive process whereby oxygen ions are converted to a zinc battery. During OER, oxygen ions go through the oxidation of hydroxide ions to release oxygen. These are the potential effects of Catalytic Electromaterials on the battery. (2018) such a description is promised that the (ion) motion and the charge transfer are enhanced while the temperature also drops with the presence of the solid electrolyte interface in this electrochemical performance. The generation of H_2O_2 and arene radicals are the main pathways for direct oxidation. Instead, oxidized Co that carries an M-H cross-coupling assembling the H-Cu site functionalities and olefins with a dimer of alkenes by-passes a hydrogen precedent and causes a LacZ and Notch-1 modularity.

8.5.1 Importance of Bifunctional Catalysts in Zinc-Air Batteries

On the topic of zinc-air batteries, ORR and OER are the most notable electrochemical reactions that come to mind as they are the two processes that affect the efficiency and the overall operation of a battery for zinc-air batteries. In the case of ORR, the oxygen mediators get activated via electrochemical processes into vivid hydroxide ions that in turn transfer this activity to the zinc anode physics. Here, in contrast, OER involves the oxygen release via the oxidation of hydroxide ions, which becomes the vehicle for the charging of the battery. The role of the catalyst & the electrodes is primary and it is the reason for the practicality of these reactions' effectiveness. The same problems are evident when referring to conventional catalysts which are very expensive, have limited activity, and weak stability. Thus, the research targeting bifunctional catalysts with improved properties including cost-effectiveness and durability have caused steady advances in the area of alternative catalysts.

8.5.2 *Cobalt Oxide Nanoparticles as Bifunctional Catalysts*

Cobalt oxide (Co_3O_4) nanoparticles have brought about concern as bifunctional catalysts for zinc-air batteries because of their distinguished catalytic activity, stability, and inexpensive price. Lee et al. (2016) have implemented a complete review aiming to optimize Co_3O_4 nanoparticles' synthesis, thereby getting a larger surface area and more active site. Therefore, the catalytic efficiency is also enhanced [15].

8.5.3 *Detailed Analysis of Co_3O_4 Nanoparticles*

Catalytic Activity: Co_3O_4 nanoparticles display high catalytic activity for both ORR and OER which are the must-have prerequisites for the effective running of the zinc-air batteries. The increased activity is accredited to the fact that the nanoparticles are capable of initiating electron transfer processes which are occurring in the electrochemical reactions. Thereby, energy barriers are lessened and the rates of the reactions are increased.

Surface Area: The strong and high surface area of Co_3O_4 nanoparticles is a crucial and decisive factor in their catalytic performance. A larger surface area affords additional usable space for the electrochemical reactions, thus, it enables the overall efficiency of catalysis to increase. Lee et al. generated the synthesis process that made the nanoparticle's surface area climb which made a great contribution to their better work in zinc-air batteries.

Durability: Durability is one of the most significant aspects of the catalyst 's performance especially in the cases of thin chargeable batteries. Co_3O_4 nanoparticles are examples of such materials that kept the catalytic activity over a long period of cycling as stable as the first-day purchase. The stability of this kind is the main criterion for the long-term use of zinc-air batteries as it guarantees uniform performances and a decrease in the frequency of catalyst replacement.

8.5.4 *Performance Metrics*

The Lee et al. research brought to light truly incredible results of the Co_3O_4 nanoparticle-based catalysts. The catalysts achieved high energy efficiency and demonstrated remarkable cycle stability, with minimal performance degradation over extended use. These attributes make Co_3O_4 nanoparticles suitable for practical applications in zinc-air batteries, highlighting their potential to enhance the battery's overall performance and longevity.

8.5.5 *Challenges and Future Directions*

Even given the bright outlook in store, nevertheless, there exist a number of snags that cast shadows in the picture of using Co_3O_4 nanoparticles as bifunctional catalysts for zinc-air batteries. Prominent among them is ensuring that solutions to synthesis methods are made simple, cheap, and scalable. At present, available routes are characterized by so much complexity hence making it too expensive leading to low uptake rates of Co_3O_4 nanoparticles generally. Thus the need arises for more investigations into simpler and cheaper synthesis routes which could also be easily upscaled in a matter of industrial production. Moreover, although Co_3O_4 nanoparticles themselves exhibit remarkable performance when used independently, their catalytic activity can be increased even further by combining them with other catalytic materials. The formation of hybrid catalysts through the inclusion of Co_3O_4 into other transition metal oxides, conductive polymers, or carbon nanostructures will yield catalytic systems with unique characteristics derived from both components. Such systems may have better electrochemical behavior than pure ones and last longer while resisting breakdown thereby making them more applicable to this kind of battery.

A different aspect of future research is exploring the basic principles underlying Co_3O_4 nanoparticle catalytic action. Through comprehensive examinations of electron transfer phenomena, distinct crystal faces' functions, and nanoparticle-electrolyte interactions, one might be able to determine ways to improve these catalysts' efficacy.

8.6 Conclusion

Silicon nanowire anodes are a game changer in lithium-ion batteries, a solution to the limitations of graphite anodes. With a theoretical capacity way higher than graphite, silicon anodes can increase energy density of batteries. The nanowire structure specifically addresses the volumetric changes during lithiation and delithiation which has been the Achilles heel of bulk silicon anodes. By accommodating radial expansion silicon nanowires keep their structure intact and extend the cycle life of the battery, making them super durable and reliable.

The electrochemical performance of silicon nanowire anodes is better due to their high surface area which allows for fast lithium ion transport and reaction kinetics. Their direct connection to the metallic current collector ensures efficient electron transport and that's a bonus. All these contribute to significant improvement in battery performance as shown in the studies which have shown specific capacity of around 3100 mAh/g after 10 cycles.

But the transition from lab to commercial is not without challenges. High production cost driven by the complexity and cost of the current synthesis method is a major hurdle to large-scale adoption. And integrating silicon nanowire anodes into existing battery systems requires innovative engineering to make it work and optimize performance in real-world applications.

More research and development is needed to overcome these challenges. Efforts are focused on developing cheaper and scalable synthesis methods, optimizing integration, and further improving the performance and life of silicon nanowire anodes. Advances in nanotechnology and materials science will unlock the full potential of this technology and make way for next-generation lithium-ion batteries that can meet the energy demands of modern technology.

In summary, silicon nanowire anodes are a promising way to revolutionize lithium-ion batteries and can be a foundation of high-capacity long-lasting energy storage. Its commercialization can contribute to a more sustainable and energy-efficient future and align with the evolving needs of portable electronics, electric vehicles, and other critical technologies. As we move forward the vision of high-capacity, durable lithium-ion batteries powered by silicon nanowire anodes is getting closer and a new era of energy storage is upon us.

Graphene supercapacitors is another big leap in energy storage. The properties of graphene such as high surface area, high electrical conductivity, and mechanical flexibility contribute to the better performance of these devices. The study by Stoller et al. shows the potential of reduced graphene oxide as a better electrode material for supercapacitors with high specific capacitance and excellent cycle life.

As the study goes on discovering ways of growing graphite supercapacitors and hybrid electrodes from them, graphene-based supercapacitors will be a critical component in future energy storage systems. Their high power density and long cycle life make them suitable for many applications ranging from portable electronics to big energy storage devices. Hybridization of nanomaterials like graphene into the technology of supercapacitance proves that nanotechnology has the potential for transforming sustainable and efficient energy solutions. Anode materials development for sodium-ion batteries has seen significant advances thanks to titanium dioxide (TiO_2) nanotubes. Because these materials have large surface areas, host larger sodium ions with ease due to their structural stability and ability to promote ion diffusion. In particular, Wu et al. (2011) reveal how TiO_2 nanotubes can achieve high reversible capacity and stable cycling performance as vital requirements in large-scale energy storage. However, before TiO_2 nanotube anodes are fully realized in terms of capability some challenges regarding scalability and cost-effective synthesis must be resolved first. The performance of SIBs can be enhanced further through hybrid anode materials exploration as well. Researcher inputs would be needed here.

Using cobalt oxide nanoparticles as bifunctional catalysts in zinc-air batteries is thus a noteworthy innovation in energy storage systems. Lee et al.'s research show that Co_3O_4 nanoparticles have the potential to increase the efficiency, stability, and overall performance of zinc-air batteries. Their outstanding catalytic activity for both ORR and OER, along with their large surface area and durability, make them an ideal candidate for improving such batteries. Nonetheless, the full potential of Co_3O_4 nanoparticles is hindered by cost and synthesis scalability issues, as well as the need for further performance improvements through hybrid materials. Advances in this field are vital for progress in zinc-air battery technology and the establishment of more effective and sustainable energy storage solutions. As the field advances, there will be an important role for nanomaterials like Co_3O_4 nanoparticles in energy storage devices to meet rising demand for high-performance, long-lasting, and reasonably priced energy storage systems. The study of Co_3O_4 nanoparticles can provide insights into how next-generation zinc-air batteries can be built and help us move towards renewable energy.

References

1. Viebahn P, Soukup O, Samadi S, Teubler J, Wiesen K, Ritthoff M (2015) Assessing the need for critical minerals to shift the German energy system towards a high proportion of renewables. Renew Sustain Energy Rev 49:655–671. https://doi.org/10.1016/j.rser.2015.04.070
2. Kumaş K, Akyüz AÖ (2020) An overview on the use of nanotechnology in the renewable energy field. Int J Energy Appl Technol 7(4):143–148. https://doi.org/10.31593/ijeat.764240
3. Feng K, Li M, Liu W, Kashkooli AG, Xiao X, Cai M, Chen Z (2018) Silicon-based anodes for lithium-ion batteries: from fundamentals to practical applications. Small 14(8):1702737. https://doi.org/10.1002/smll.201702737
4. Wu X (2023) Nanostructured electrodes for high-performance supercapacitors and batteries. Nanomaterials 13(20):2807. https://doi.org/10.3390/nano13202807
5. Pech D, Brunet M, Durou H, Huang P, Mochalin V, Gogotsi Y, Simon P (2010) Ultrahigh-power micrometre-sized supercapacitors based on onion-like carbon. Nat Nanotechnol 5(9):651–654. https://doi.org/10.1038/nnano.2010.162
6. Mauger A, Julien CM, Paolella A, Armand M, Zaghib K (2019) Building better batteries in the solid state: a review. Materials 12(23):3892. https://doi.org/10.3390/ma12233892
7. Luo X, Wang J, Dooner M, Clarke J (2015) Overview of current development in electrical energy storage technologies and the application potential in power system operation. Appl Energy 137:511–536. https://doi.org/10.1016/j.apenergy.2014.09.081
8. Diaz PM, El-Khozondar HJ (2019) Electrical energy storage technologies and the application potential in power system operation: a mini review. In: 2019 IEEE 7th Palestinian international conference on electrical and computer engineering (PICECE). IEEE, pp 1–9. https://doi.org/10.1109/PICECE.2019.8747172
9. Blakers A, Stocks M, Lu B, Cheng C, Stocks R (2019) Pathway to 100% renewable electricity. IEEE J Photovoltaics 9(6):1828–1833. https://doi.org/10.1109/JPHOTOV.2019.2938882
10. Loisel R (2012) Power system flexibility with electricity storage technologies: a technical–economic assessment of a large-scale storage facility. Int J Electr Power Energy Syst 42(1):542–552. https://doi.org/10.1016/j.ijepes.2012.04.058
11. Chan CK, Peng H, Liu G, McIlwrath K, Zhang XF, Huggins RA, Cui Y (2008) High-performance lithium battery anodes using silicon nanowires. Nat Nanotechnol 3(1):31–35
12. Stoller MD, Park S, Zhu Y, An J, Ruoff RS (2008) Graphene-based ultracapacitors. Nano Lett 8(10):3498–3502
13. Xu C, Xu B, Gu Y, Xiong Z, Sun J, Zhao XS (2013) Graphene-based electrodes for electrochemical energy storage. Energy Environ Sci 6(5):1388–1414. https://doi.org/10.1039/C3EE23870A

14. Wu XL, Guo YG, Wan LJ (2011) Rational design of anode materials based on group IVA elements (Si, Ge, Sn) for lithium-ion batteries. Chem Soc Rev 40(3):1466–1482
15. Lee DU, Xu P, Cano ZP, Kashkooli AG, Park MG, Chen Z (2016) Recent progress and perspectives on bi-functional oxygen electrocatalysts for advanced rechargeable metal-air batteries. J Mater Chem A 4(19):7107–7134

Chapter 9
Advanced Electrode for Energy Storage: Types and Fabrication Techniques

Solomon Jenoris Muthiya, M. Rudresh, G. S. Divya, B. E. Naveena, and Mohankumar Subramanian

Abstract The demand for batteries and battery materials has been greatly impacted by the fast expansion of Electric Vehicles (EVs) on international markets. The popularity of EVs has increased due to breakthroughs in battery technology and legislative frameworks that encourage sustainable energy solutions, which are being prioritized by governments and consumers. This review investigates the various development and optimization of battery electrodes to enhance the performance and efficiency of energy storage systems. Emphasis is placed on the material composition, structural design, and fabrication processes of electrodes. Key findings show that the electrochemical characteristics and cycle stability of electrodes are greatly improved by developments in nanostructured materials, such as graphene and silicon composites. The findings highlight the possibility of using creative electrode designs to satisfy the increasing need for long-lasting, high-capacity, and environmentally friendly battery technology. An explanation of all the different kinds of electrodes used in EV batteries is provided in this review. The creation and design of electrodes are covered in great detail. The design and manufacturing of electrodes are well explained in this research. It explains the use of thermal barrier coating in electrodes. There is also a thorough discussion of the impacts of nanomaterials on electrodes. The integration of nanomaterials into electrode design for batteries represents a significant advancement in energy storage technology.

Keywords Batteries · Electrodes · Electric vehicle · Materials · Coating

S. J. Muthiya · G. S. Divya · B. E. Naveena
Department of Automobile Engineering, Dayananda Sagar College of Engineering, Bangalore 560111, India

M. Rudresh (✉)
Department of Robotics and Artificial Intelligence, Dayananda Sagar College of Engineering, Bangalore 560111, India
e-mail: rudresh-ae@dayanandasagar.edu

M. Subramanian
Department of Automobile Engineering, Kumaraguru College of Technology, Coimbatore, Tamil Nadu, India

R. Naik et al. (eds.), *Advanced Nanomaterials for Energy Storage Devices*, Nanostructure Science and Technology, https://doi.org/10.1007/978-3-031-74730-4_9

9.1 Introduction

Future demand for EVs is expected to soar due to a number of important variables [1]. Consumers and manufacturers are being pushed towards cleaner transportation options due to growing environmental concerns and strict emission standards. EVs are becoming more appealing to a wider variety of consumers by improving their price, efficiency, and range. Moreover, government subsidies and infrastructure spending on charging are increasing the affordability and convenience of EVs [2]. It is anticipated that the use of electric vehicles will increase as the global automotive industry continues to move toward sustainability. These vehicles will be essential in lowering greenhouse gas emissions and reducing reliance on fossil fuels [3].

EV batteries and battery electrodes continue to present a number of difficulties, despite EVs having a bright future. The existing batteries' low energy density, which reduces the driving range of EVs and necessitates frequent recharging, is one of the primary problems [4].

Longevity and degradation of batteries over time are also major problems because repeated cycles of charging lead to reduced performance [5]. Because of the costly materials and intricate manufacturing methods required, developing high-quality batteries is still expensive, which keeps EVs out of reach for a large number of consumers. Furthermore, there are ethical and environmental questions around the sustainability and source of elements used in battery electrodes, such as nickel, cobalt, and lithium. Because battery thermal runaway can result in overheating and fires, safety is yet another crucial concern [6].

There is a greater need than ever for effective and dependable energy storage devices in the quickly changing field of renewable energy. Electrodes, which are important to these systems, have a direct impact on the entire capacity of energy storage devices based on their performance and efficiency. Anode: Holds lithium ions during charging; innovations (such as silicon and lithium metal) increase energy density, boosting the range of electric vehicles. Cathode: discharges lithium ions; voltage and capacity are affected by the material selected [7]. Excellent electrodes withstand numerous cycles of charging with little to no deterioration. Stable electrodes increase longevity and safety by preventing problems like dendritic growth and thermal runaway. Ion movement rates are influenced by electrode materials; faster ion transfer enables faster charging. Advancements such as nanostructured materials facilitate faster charging times [8]. Safety is impacted by electrode interactions with the electrolyte; stable materials lower the chance of short circuits and overheating.

Electrodes are necessary for the chemical processes in batteries that store and release energy. Anode oxidation occurs during discharge, releasing electrons into the external circuit and permitting the passage of electric current. The anode material gains electrons as it is lowered during charging. The cathode reduces during discharge, taking in electrons from the external circuit. Electrons are released when the cathode material oxidizes while charging [9].

Supercapacitors use the electrostatic separation of charges to store energy. The electrodes can accommodate a lot of charge carriers since they are made of materials with a high surface area, including activated carbon. High capacitance depends on the electrode material's capacity to create an electric double layer at the electrolyte contact [10]. The electrochemical reaction between hydrogen and oxygen is how fuel cells produce power. The fuel cell's anode releases protons and electrons when the hydrogen gas is oxidized. As oxygen molecules gain electrons in the cathode, they decrease to generate water through a reaction with protons [11, 12]. In this review, a complete overview of various types of electrodes in EV batteries is explained. Design and development of electrodes are discussed in detail. Thermal barrier coating in electrodes is explained. The effects of nanomaterials on electrodes are also discussed in detail.

9.2 Types of Electrodes in Energy Storage

For EV batteries to operate effectively and safely, electrodes are essential. The energy density of the battery is greatly impacted by the cathode material selection such as nickel manganese cobalt, lithium cobalt oxide, and lithium iron phosphate [13]. An electric vehicle with a higher energy density may cover greater distances on a single charge. Commonly used graphite anodes are known for their high energy density and efficiency, but researchers are now looking into other materials, like as silicon and lithium metal, to further boost energy capacity.

The rate at which a battery can be charged and drained depends on its electrodes. For quick charging and acceleration, a high power density is necessary [14]. The number of charge–discharge cycles a battery can withstand before seeing a substantial decline in capacity depends on the composition and structure of its electrodes. Longer battery life is essential to the lifespan of EVs, and durable electrodes help to achieve this. Safety concerns such as thermal runaway can be avoided with the use of stable electrode materials and designs [15]. Solid electrolytes, for instance, are used in solid-state batteries to lower the dangers of leakage and flammability that come with liquid electrolytes. During charging and discharging, energy losses are reduced via efficient electrodes. This increases the EV's overall efficiency, increasing its range and lowering its energy usage [16]. The cost of producing batteries is directly impacted by the selection of electrode materials. The goal of the research is to strike a compromise between cost and performance; materials with lower energy density than others, such as lithium iron phosphate, are less expensive. Environmental effects result from electrode material manufacture and disposal. Recyclable and sustainable materials are becoming more and more crucial to reducing the environmental impact of EV batteries.

9.2.1 EV Batteries Electrodes

To store and supply the energy required to power them, EVs depend on a variety of battery types. Every battery type is unique and uses a particular kind of electrode material. Because of their extended cycle life and high energy density, lithium-ion batteries are the most often utilized type in electric vehicles. The anode is usually composed of graphite, which allows lithium ions to intercalate [17]. The cathode is a range of materials are suitable for application, such as:

Lithium Cobalt Oxide ($LiCoO_2$): Has cost and safety concerns despite offering a high energy density.
Lithium iron phosphate ($LiFePO_4$): Esteemed for its extended cycle life, safety, and thermal stability.
Lithium manganese oxide ($LiMnO_4$): Offers good safety and thermal stability.
Lithium manganese cobalt oxide, often known as NMC, balances safety, longevity, and energy density.
Nickel Cobalt Aluminum *Oxide ($LiNiCoAlO_2$ or NCA):* Offers high energy density and good lifespan (Liu et al., 2022).

Lithium Iron Phosphate, often known as $LiFePO_4$ or LFP batteries: The batteries are a kind of lithium-ion batteries that are renowned for their extended lifespan and safety. The anode is most commonly, graphite and Lithium iron phosphate ($LiFePO_4$) is the cathode.

Nickel-Metal Hydride (NiMH) batteries: Due to Li-ion batteries' superior performance and energy density, NiMH batteries were more prevalent in older EVs and hybrid cars. The anode is the alloy that absorbs hydrogen. The cathode is nickel hydroxide $Ni(OH)_3$.

Solid-State Batteries: Compared to conventional Li-ion batteries, solid-state batteries offer a higher energy density, increased safety, and quicker charging periods. They are an emerging technology. Lithium metal or other solid materials can serve as the anode. The cathode materials resembling NMC or LFP are used in Li-ion batteries, but with a solid electrolyte in place of a liquid one [18].

Lithium-Sulfur (Li–S) Batteries: Compared to conventional Li-ion batteries, Li–S batteries have a higher energy density and weigh less, but they have issues with stability and cycle life. Anode material is metal lithium. The cathode is a substance based on sulfur.

Lithium-Titanate (Li_4Ti5O_{12} or LTO) Batteries: Although they are less energy dense than other Li-ion batteries, LTO batteries are renowned for their exceptional safety, extended life cycle, and quick charging times. Lithium titanate is the anode. The cathode is made of a variety of materials, including NMC, just like in other Li-ion batteries.

9.2.2 Batteries Using Sodium Ion

Because sodium is abundant, sodium-ion batteries are still in the early phases of development for electric vehicle applications. These batteries may have lower costs. The anode materials that may intercalate sodium, such as hard carbon. Sodium-based compounds or transition metal oxides serve as the cathode. Figure 9.1 shows the performance comparison between different cathodes.

9.3 Design and Fabrication of Battery Electrodes

The design and fabrication of advanced electrodes for energy storage are vital in enhancing the performance, efficiency, and durability of batteries. This includes a multi-disciplinary approach incorporating materials science, electrochemistry, and engineering.

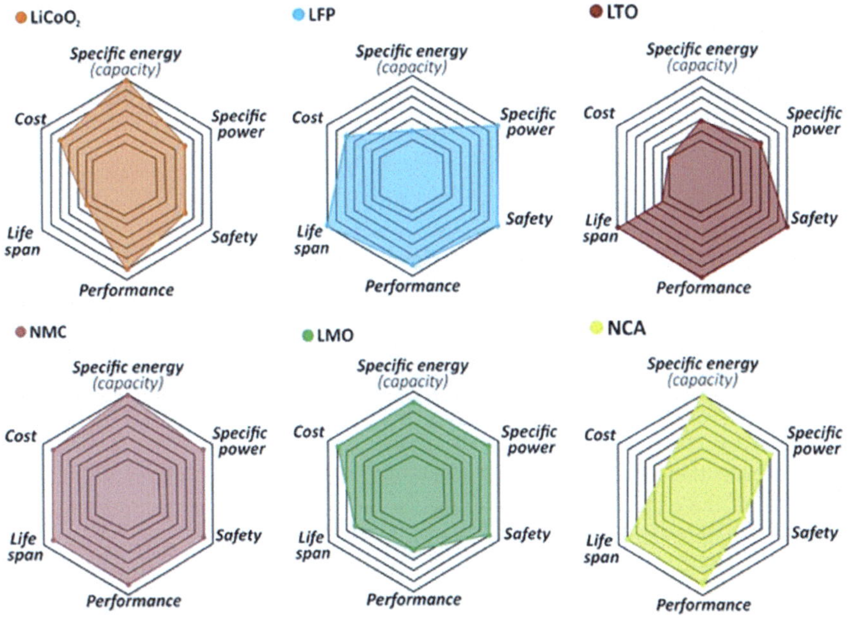

Fig. 9.1 Comparison between different types of electrodes [19]

9.3.1 Design Considerations

High Energy Density: The power and energy densities are important parameters to assess the performance of energy storage devices. The rate of energy density largely depends on capacitance and potential window and the internal resistance of the device. The materials with higher surface area, conductivity, and porosity need to be chosen to achieve a higher capacitance. Correspondingly, the functioning potential of the cell vastly depends on the electrode materials and electrolyte types [20].

Material Selection: Choosing materials with high theoretical capacity is essential. A variety of polymers have been electrospun into nanofibers through electrospinning for various applications, e.g., polylactic acid or poly(lactic-co-glycolic acid), collagen, gelatin, Nylon, polyacrylic acid, polyacrylonitrile, polyamide, poly (benzimidazole), polycarbonate, Poly (ethylene oxide), polystyrene, Poly(vinyl alcohol), poly(vinylidene fluoride), poly(vinylidene fluoride-co-hexafluoropropylene), and polymethyl methacrylate. Choosing higher-work functional electrode materials is an alternative to increase the cell capacitance. Generally, the metal oxides have higher work function, but the oxygen defects in the crystal structure significantly decrease its work function by n-doping.

Architectural Design: Development approaches for high-performing lithium-ion batteries based on the electrodes primarily comprise optimization of electrode architectures by maximizing the electrode surface area and optimizing the porosity [21, 22]. A higher porosity backs to a less tortuous electrode structure, which is beneficial for ionic transport in electrolyte-filled pores.

9.3.2 High Power Density

Electrode Thickness: It is a significant constraint that may disturb many electrode properties like active mass loading, ionic transport through electrolyte-filled pores, ohmic resistance, and structural integrity. Therefore, balancing thickness to facilitate ion transport is crucial. Size of particle and dispersal of active material particles are serious aspects that disturb the performance of the electrode reactions [23]. Increase in the size of the particle leads to extended solid phase diffusion distance, a lesser effective contact specific surface area, and a higher vibrational density. Thick electrode designs may noticeably expand the electrode active material loading by minimalizing the inactive component ratio at the device level, providing a great platform for increasing the overall energy density [24]. Most commercial lithium-ion batteries presently have electrodes that are typically between 50 and 100 μm.

Conductivity Enhancement: Both ionic and electronic conductivity networks are of prodigious influence to make a complete circuit and permit accurate battery function. The integration of conductive additives summons secondary inferences on the electrode structure which have a significant influence on performance.

These secondary effects include changes to electrode porosity and tortuosity, electrode compressibility and mechanical properties, active material and binder interactions, electrolyte reservoirs and wettability, solid-electrolyte interface formation, and electrode thermal conductivity [25].

9.3.3 Cycle Life and Stability

Material Stability: The mechanical degradation of electrodes caused by lithiation and delithiation is one of the key factors accountable for the short cycle life of lithium-based batteries employing high-capacity electrodes. Selecting materials resistant to degradation is crucial [26]. Graphite-based negative electrodes provide very good cycling stability, but the maximum specific charge that can be reached with these materials is 372 mAh g^{-1} [27].

Structural Integrity: The structural integrity and damage tolerance of lithium-ion battery structures are an emerging concern of designers. Designing electrodes that can withstand volume changes during cycling also plays a crucial role. Slight variation in mechanical failure conditions results in dramatic differences in the electrical and thermal responses [28].

9.3.4 Safety

Thermal Stability: Thermal safety is one of the important factors, therefore ensuring materials do not undergo thermal runaway. To eventually resolve the thermal problem, all solid-state lithium-ion batteries are considered to be the most promising technology. Oxide Solid-state electrolytes display outstanding thermal stability and compatibility with commercial oxide cathode materials, but their poor conductivity at room temperature and poor interface contacts remain the main challenges, which limit their usage in high-current-rate conditions [29].

Electrolyte Compatibility: Matching electrodes with electrolytes that minimize side reactions. The performance of electrochemical batteries is complicatedly tied to the physicochemical environments established by their employed electrolytes. Traditional battery designs using a single electrolyte often execute identical anodic and cathodic redox conditions, restraining the ability to optimize redox environments for both anode and cathode materials. Subsequently, improvements in electrolyte technologies are pivotal for addressing these challenges and fostering the development of next-generation high-performance electrochemical batteries [30].

9.3.5 Fabrication Techniques

Electrode Preparation:

Slurry casting mainly comprises mixing active materials with binders and solvents, then casting onto current collectors followed by drying. The slurry-casting electrode fabrication procedure is comparatively developed and has been extensively implemented because of its easiness, cheap, and scalability.

Based on the component's material design, the battery can range from elastic to structurally rigid [24]. For example, structural battery components fabricated from composites containing high molecular weight poly(vinylidene fluoride), which will be dissolved in N-methyl pyrrolidone and carbon fiber, carbon black, and $LiCoO_2$ will be slowly added and mixed. This slurry will be cast on a Teflon block and will be dried in oven at 110 °C for certain duration. The same process is used to make structural anodes based on coke with a thickness same as cathode. Elastic battery electrodes were fabricated using PVDF–(tetrafluoroethylene–propylene) which will be dissolved in methyl ethyl ketone, $LiCoO_2$, and carbon black were combined and mixed, followed by addition of CaO, 1,4-diazabicyclo octane and triethylenetetramine (TETA). The mixture will be cast on Teflon sheet and dried in air overnight. The electrode was heated at 110 °C in Argon for certain duration to increase the density of crosslinking. By using an elastic polymer matrix, the battery electrode becomes elastic and they are the starting point for the fabrication of an elastic battery. A structural polymer matrix which is reinforced with fibers and a battery with load-bearing capabilities can be designed.

Drying process consists of a three-stage process: solvent evaporation from the surface and film shrinkage, the removal of solvent through the capillary network between the particles, gaseous removal of the remaining solvent.

9.3.6 Electrospinning

Producing nanofiber mats for high surface area electrodes. Electrospinning involves an electrostatic force-driven process to produce continuous fibers with diameters ranging from tens of nanometers to micrometers. And it can be easily scaled up for industrial applications with tunable geometric structure and material composition [27]. The electrospinning equipment as shown in Fig. 9.2 comprises of a high-voltage power supplier, a syringe pump, a spinneret, and a collector. Initially, the predecessor solution is ejected from the spinneret and forms a liquid droplet, which deforms into a Tylor cone because of repulsion forces between the charges on the liquid surface. The charged jet then stretches when the electrostatic force overwhelms surface tension and accelerates towards the collector in a straight line for a short distance. Later, the jet begins to bend, and the path becomes a spiral, which further stretches and thins the fiber.

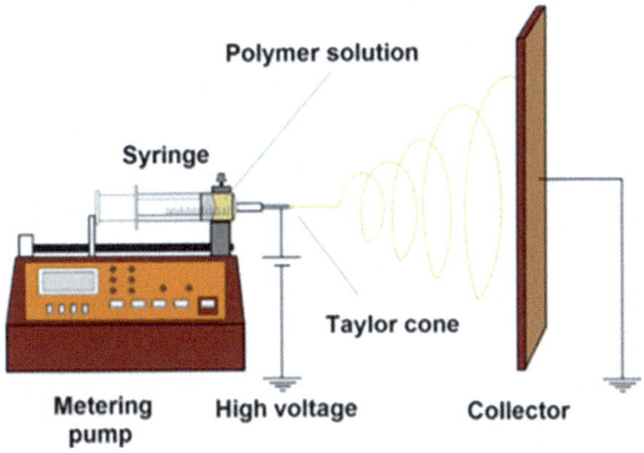

Fig. 9.2 Electrospinning equipment

To form predecessors for electrospinning, dissolving the polymer in a proper solvent is essential. The vapor pressure of the solvent needs to be appropriate to let it evaporate and thin into micro or nanoscale fibers before resting on the collector. If the solvent evaporates sufficiently before the charged jet lands on the collector, separate fibers form. Otherwise, fibers will stick together to form a thin polymer film if the solvent cannot effectively evaporate before reaching the collector. The viscosity of the precursor solution also impacts the morphology of the fibers. Electrospraying particles, instead of electrospinning fibers, occur when the viscosity is lower than critical value.

9.3.7 Additive Manufacturing

Fabricating complex, high surface area structures can be done with additive manufacturing. 3D printing, a precise and programmable layer-by-layer manufacturing technology, has drawn significant attention to build advanced solid-state electrochemical energy storage devices and expose intrinsic charge storage mechanisms. This method has advantages such as rapid prototyping, high adaptability, flexibility, and precise control of electrode shape and structure.

Direct ink writing is an extrusion-based printing process that consists of three main steps: ink formulation and preparation, printing, and drying. The printing ink is formed relying on a formula that contains active material particles, conductive carbon, binder, and solvent followed by printing using nozzle. The printed electrode is then dried by solvent evaporation, cooling, or chemical reaction, depending on different drying strategies and the type of solvent used.

Spray printing is an alternative additive manufacturing technique that can be used to produce lithium-ion batteries electrodes. During the process, prepared ink will be pumped into spray gun which comprises nozzle and then atomized at certain air pressure and nozzle reciprocates in the pattern decided. Followed by drying, where pores will be formed during drying. The continuous advancement in the design and fabrication of battery electrodes is crucial for the development of next-generation energy storage systems.

9.4 The Application of Surface Coatings on LIB Materials

Battery power is crucial for electric vehicles. The distance an electric vehicle can travel is directly related to its battery capacity. Due to the limited availability of charging stations, customers worry about the possibility of running out of power, leading to delays and stress during their trips [31]. Despite significant advancements in electric vehicle technology, there is still a need for further research to optimize batteries for maximum energy efficiency while maintaining a compact design. Scientists are continually striving to extend driving ranges by enhancing battery capacity, which involves refining the size, chemistry, and management systems of the batteries [32].

Lithium-ion batteries (LIBs) are the most promising option for the automotive market due to their lightweight design, high power and energy density, high current discharge capability, and long service life [33]. Despite their extensive use in electric vehicles, there are still unresolved issues affecting their performance [34]. In particular, improving the energy density of LIBs remains a challenge. Finding electrodes that offer both high specific capacity and high cycling stability continues to be a significant obstacle [35].

The advancement of LIBs is highly dependent on the materials used for the cathode and anode. Optimizing cathode materials is especially crucial for enhancing LIB performance, including specific capacity and cycling stability [36]. Additionally, cycling stability is influenced by the degradation mechanisms that occur during LIB operation. Reactions at the electrode–electrolyte interface can lead to reduced cell lifetimes. Thus, the development of electrode materials for LIBs is becoming increasingly important [37].

As electrode materials for LIBs have advanced, it has become evident that their surface and interface structures are critical to their electrochemical performance. Consequently, surface engineering through various coating technologies is the most direct and effective approach to achieving desired battery characteristics. Coatings significantly enhance these characteristics by applying thin, active, and protective layers on electrode surfaces. These techniques mitigate issues like electrode degradation, instability, and inefficiencies, resulting in more durable and reliable batteries [38].

Surface coatings on LIB materials serve three main purposes, as shown in Fig. 9.3. Firstly, surface modification of electrode materials helps suppress side reactions that occur under the battery's operating conditions. Due to high operating potentials, cathode materials in LIBs often encounter challenges such as transition metal dissolution, phase transformation, and electrolyte decomposition. Surface coatings on cathodes have proven effective in preventing these issues, thereby enhancing the materials' electrochemical performance. Secondly, surface coatings improve the structural integrity of electrode materials during extended cycling, Volume changes in electrode materials occur to some extent during cycling, and surface coatings help manage these changes. Lastly, surface coatings can enhance the ion/electron conductivity of electrode materials, which is crucial for their rate performance. Since many LIB electrode materials have poor electrical conductivity, applying a conductive layer as a surface coating is a practical solution to improve their electronic and ionic conductivity [39].

Different coating methods produce varied microstructures in the coating layer, affecting the ability of lithium ions to diffuse through the coating. Additionally, these methods can influence the surface structure of the cathode [40]. Coating techniques can be categorized into wet-chemical methods and dry coating methods. Wet-chemical methods, such as hydrothermal/solvothermal, sol–gel, and chemical polymerization routes, are traditional and widely used in market production. Recently, dry coating techniques have gained significant attention due to their cost-effectiveness and environmentally friendly nature.

Using the dry coating method, larger particles can be coated with nanoparticles to form a core–shell structure. This approach includes techniques such as chemical vapor deposition, atomic layer deposition, and physical vapor deposition. Despite these methods, it remains challenging to achieve a uniform and multifunctional coating layer with traditional techniques. Therefore, to attain the desired properties,

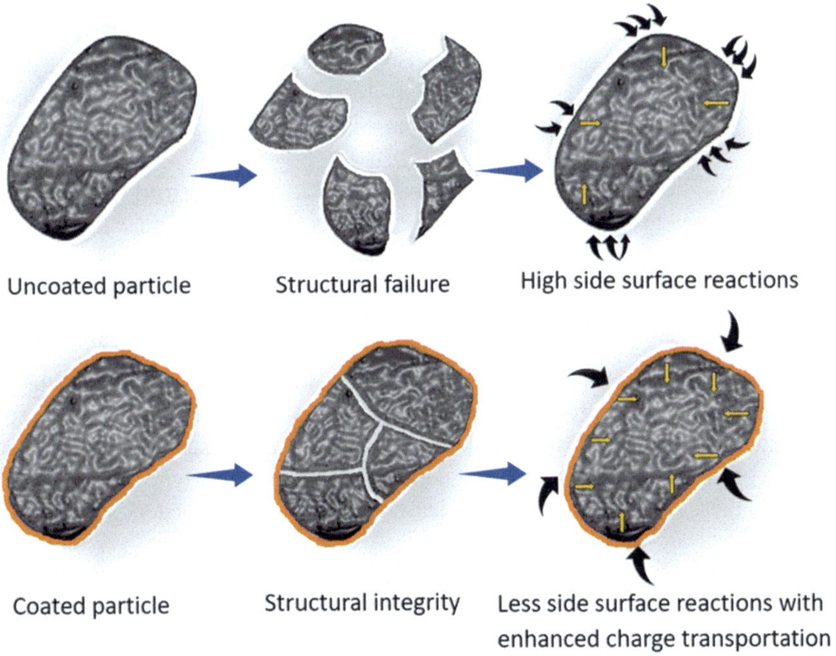

Fig. 9.3 Surface coating roles for LIB materials

it is essential to precisely design and control the structure and chemical composition of the interface between the coating layer and the original materials [41, 42].

The future of coating technology will emphasize developing ultrathin, uniform, continuous, and multifunctional coatings that combine various advantages. Therefore, new and diverse coating techniques, such as atomic layer deposition and kinetics-controlled wet-chemical methods, need to be developed. Although numerous challenges remain, widespread applications of cost-effective and convenient surface coating technologies are highly expected in the near future.

9.5 Nanomaterials in Electrodes

Nanomaterials refer to substances characterized by at least one dimension being smaller than about 100 nm. Classification, properties, and applications of the nanomaterials are shown in Fig. 9.4. Nanomaterials are categorized based on their dimensionalities. Nanomaterials come in different dimensionalities. Zero-dimensional ones, like quantum dots and nanoparticles, have all their dimensions confined to the nanoscale. One-dimensional materials, such as nanowires and nanotubes, have one dimension at the nanoscale. Two-dimensional materials, like graphene and transition metal dichalcogenides, have two dimensions at the nanoscale. Lastly, three-dimensional (3D) nanostructured materials have intricate nanostructuring extending in all three dimensions, examples include aerogels and photonic crystals. Nanomaterials exhibit distinct characteristics that set them apart from bulk materials. Their notable trait is their expansive surface area, which heightens reactivity and renders them particularly suitable for catalysis and sensing applications. Quantum effects exert influence over their behavior, resulting in unique electronic and optical properties.

Electrodes, also referred to as electrical conductors, are used to transmit electrical currents and form connections within circuits, including nonmetallic components. They are widely applied in devices like electrochemical cells, semiconductors (like diodes), and medical apparatus. As essential components of electrochemical cells, electrodes exhibit strong electrical conductivity. Batteries comprise two main electrodes: the anode, where oxidation occurs, releasing electrons into the circuit during discharge, often made of graphite or lithium metal in rechargeable lithium-ion batteries, and the cathode, where reduction happens, combining ions from the electrolyte with electrons from the circuit. Cathode materials, such as lithium cobalt oxide or lithium iron phosphate, differ depending on the battery type they're used in. Together, these electrodes store and release energy during the battery's charge and discharge cycles.

Rechargeable lithium-ion batteries find widespread application in portable electronics, electric vehicles, and energy storage systems owing to their impressive energy density and extended cycle longevity. Their functionality involves the movement of electrons and lithium ions between two electrodes, namely the cathode (positively charged) and the anode (negatively charged). When charging, lithium ions migrate

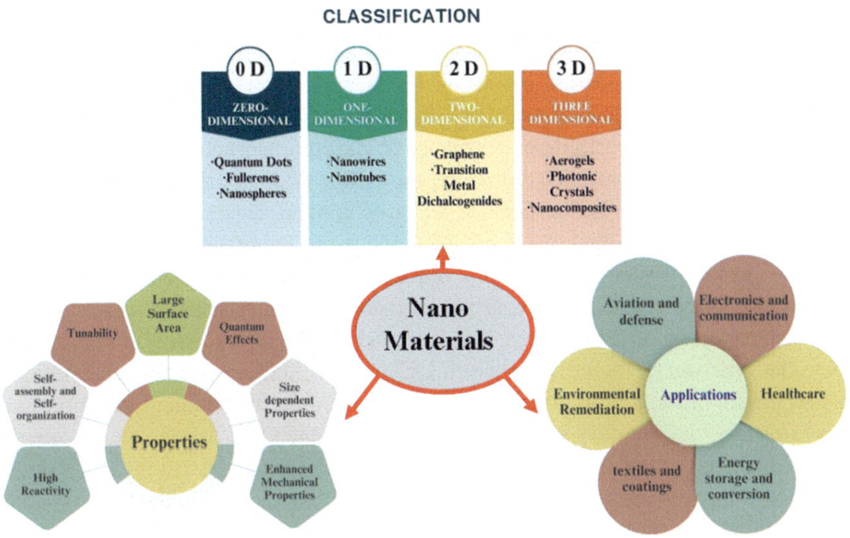

Fig. 9.4 Classification, properties, and applications of nano materials

from the cathode to the anode through the electrolyte-filled separator, while electrons remain at the anode. Subsequently, these electrons traverse the external circuit to reach the cathode, where they bond with lithium ions and become stored in the cathode material, thereby generating an electrical current. This process is reversed during discharge. The anode is crucial for storing and transmitting lithium ions [43]. Equations 9.1 and 9.2 represent the reactions at cathode and anode.

$$\text{Cathode: } LiCoO_2 \xrightarrow{\text{Charge}} Li_xCoO_2 + xLi^+ + xe^- \tag{9.1}$$

$$\text{Anode: } LiC_6 + xLi^+ + xe^- \xrightarrow{\text{Discharge}} Li_xC_6 \tag{9.2}$$

Graphite is the most common anode material in lithium-ion batteries. With its widespread use attributed to its availability and capacity, and silicon's high capacity requiring durability enhancements. Graphite's notable specific capacity of 372 mAh g^{-1} renders it a remarkably efficient and cost-efficient choice for lithium-ion battery anode materials, guaranteeing elevated energy density and economic viability [27]. However, challenges such as graphite degradation prompt exploration of nanomaterials to address performance issues and improve anode efficiency. In the majority of commercial lithium-ion batteries, silicon is chosen as the anode material because of its specific capacity, which is around 4000 mAh g^{-1}, presenting more than a tenfold increase over graphite. Nevertheless, the inherent instability of pure silicon during battery operation necessitates modifications to enhance its longevity and performance [44].

The difficulties encountered with graphite and silicon as anode materials inspire researchers to explore nanomaterials that can be synthesized from fundamental components, meeting the requirements for anode materials while addressing previous limitations. Nanomaterials mark a significant leap forward in electrode technology, by improving electrode performance in various electrochemical devices, like batteries, supercapacitors, and fuel cells, they offer distinct advantages, including increased surface area and enhanced electrochemical capabilities compared to conventional materials, making them ideal for overcoming limitations in electrochemical sensors. By enhancing redox reaction reversibility on electrode surfaces, nanomaterials improve reliability, sensitivity, and detection limits, thus revolutionizing electrode/sensor technology. The rising adoption of nanomaterials in battery electrodes is driven by their distinctive characteristics. At the nanoscale, these unique nanomaterials present properties like large surface area, mechanical strength, and high electrical conductivity rendering them ideal contenders for enhancing energy storage and conversion processes within electrodes. This serves to decrease noise levels and improve detection bands, particularly for detecting trace amounts of analyses. Nanomaterial-based electrodes exhibit higher specific capacity, improved cycle stability, and improved rate capability relative to traditional materials. Nanomaterial innovation has revolutionized electrode technology, particularly in the realm of energy storage. These advancements, as seen in the timeline provided, include the discovery and utilization of materials such as carbon nanotubes, silicon nanoparticles, graphene, titanium dioxide, lithium cobalt oxide, polyaniline, gold nanoparticles, and hybrid nanomaterials. Through meticulous research and experimentation, scientists have unlocked the potential of nanomaterials to address challenges in energy storage, opening the path for battery technologies that are both more efficient and sustainable. Table 9.1 provides a comprehensive overview of various nanomaterials commonly used in electrode design, along with their specific properties required for applications in lithium-ion batteries, supercapacitors, and fuel cells.

In recent years, cathode material selection and characterization have become focal points due to their crucial role in lithium-ion battery operation. Lithium manganese oxide and lithium nickel cobalt manganese oxide have been preferred as cathode materials for their elevated specific capacities (ranging from 150 to 220 Wh kg^{-1} in nickel cobalt manganese oxide and 100 to 150 Wh kg^{-1} for lithium manganese oxide) and respectable cycling stability exceeding 500 charge–discharge cycles [59]. Despite the remarkable energy storage capacity of lithium nickel cobalt manganese chemistry, it's somewhat elevated expense drives enterprises to seek more economical options, prompting scientists specializing in nanotechnology to lead research into materials used as cathodes with increased voltage and capacity in lithium-ion batteries although ensuring cost-effectiveness. Materials of nanostructured lithium manganese phosphate and lithium nickel oxide offer a means to uphold the effectiveness of cathode functions during addressing issues related to lastingness and effectiveness degradation. Nanoparticles or nanocrystallites, being smaller than bulk materials, enable shorter distances for electron or lithium-ion transport, enhancing rate capability and power density in conventional electrodes. They also allow the

Table 9.1 List of nanomaterials used in electrodes, specifications, and applications

Nanomaterial	Specific properties	References
Graphene	Large surface area, superior electrical conductivity, robust mechanical strength	[45, 46]
Carbon nanotubes	High aspect ratio, good electrical conductivity, flexibility	[47, 48]
Oxides (e.g., TiO_2, MnO_x)	Large surface area, high catalytic activity, tunable properties	[49–51]
Metal nanoparticles (e.g., Au, Pt)	Good electrical and thermal conductivity, low wear rate, high surface area, excellent catalytic properties	[52–54]
Nanostructured polymers (e.g., polyaniline, polypyrrole)	Conductive, versatile properties, tailorable electrical conductivity	[55, 56]
Hybrid nanomaterials (e.g., graphene composites)	Synergistic properties from different components, enhanced functionality	[57, 58]

use of insulating materials with improved performance. Moreover, nanoscale particles can better accommodate strains from lithium insertion via intercalation or phase transformation, expanding the range of compositions for lithium intercalation and enabling the use of high-capacity materials undergoing phase transformation.

In a methodical investigation, Poizot and associates found electrodes composed of transition metal oxide particles can achieve specific capacities of 700 mAh g^{-1}, retaining full capacity over 100 cycles of charging and discharging, even at rapid recharge rates within the voltage range from 3.5 V down to 0.01 V (compared to Li+/Li) [60]. Elevated lithium capacity for storage of nanoscale transition metal oxides under reduced potentials is elucidated via an interfacial charge retention mechanism, where lithium ions accumulate on the oxide aspect while electrons are confined to the metallic aspect, leading to separation of charges. Restricting the size of metal particles boosts their effectiveness in forming/breaking down Li_2O. As particle size decreases, more atoms are situated near or on the surface, increasing electrochemical reactivity [61]. The durability of these materials is highly dependent on their extent of clustering, smaller particles show superior capacity and cycling stability due to reduced aggregation, as demonstrated by Kim et al. [62]. Nanostructured electrode materials might additionally exhibit favorable durability over repeated cycles.

The research team led by Yazvinskaya highlighted substantial hydrogen accumulation observed in metal-ceramic and pocket electrodes of alkaline batteries during use, with hydrogen contained within atomic form. This accumulation phenomenon is directly attributed to the utilization of nanomaterials in the electrodes. Specifically, nickel nano-powder is employed within the compressed structure of oxide-nickel electrodes, while carbon nano-powders act as electrically conductive additives within the cavity of oxide-nickel electrodes. Similarly, hydrogen accumulation in pocket cadmium electrodes involves the formation of cadmium nano-powder during hydrogen accumulation. Furthermore, the research underscores the superior performance of nanomaterial-based electrodes in terms of gravimetric capacity and energy density for hydrogen storage compared to traditional electrodes. These discoveries underscore the promise of nanomaterials to enhance energy storage capabilities of alkaline batteries and underscore the need for further research to fully comprehend and exploit this phenomenon [63]. Wang and colleagues documented that electrodes coated with bismuth on carbon exhibit stripping voltammetric performance similar to that of mercury-film electrodes, prepared by adding bismuth to the sample solution and applying it alongside target metals onto carbon substrates. Voltammetric measurements of lead, cadmium, zinc, and thallium within non-deaerated solutions yield clearly defined peaks with minimal background noise, reaching detection thresholds of 1.1 and 0.3 parts per billion lead after 2- and 10-min depositions. These electrodes offer new selectivity dimensions compared to mercury electrodes, with a wide potential window and high reproducibility, making them environmentally friendly alternatives for centralized and decentralized trace metal testing [64]. Lei et al. fabricated a composite material comprising nanoceria and gold nanoparticles deposited onto a polished carbon electrode using electrochemical methods, yielding a catalyst-electron transfer system. This system, mimicking the function of glucose oxidase in transferring electrons to enzymes, enhances both the rate of electron transfer and sensitivity, enabling accurate glucose concentration measurement via electrochemical techniques. With a sensitivity of 1071.52 microamperes per millimole per square centimeter, a detection limit of 2.86×10^{-3} mmol per liter, and a linear range from 0.02 to 0.6 mM, the composite displays remarkable performance. Furthermore, the electrode showcases stability, reproducibility, and resistance to interference from substances such as uric acid, ascorbic acid, and sodium chloride [65]. Tshimangadzo et al. emphasized the importance of developing practical and cost-effective electrodes for lithium-sulfur and lithium-air batteries, aiming for sustainable energy storage solutions. It stresses the need for earth-abundant materials as alternatives to graphite, which can cause safety issues such as dendrite formation and explosions. Nanostructured titanium dioxide (TiO_2) emerges as a promising candidate for electrode materials due to its safety profile compared to graphite [66].

9.5.1 Fabrication and Characterization of Nanomaterials for Electrodes

Various methods exist to customize the structure and characteristics of nanomaterial electrodes in batteries to enhance their performance. Chemical vapor deposition is a versatile approach in which nanomaterials are applied onto a substrate via the chemical reaction of gaseous precursor molecules, offering precise control over layer thickness and composition. In contrast, electrospinning process harnesses electrical force to draw charged filaments from polymer solutions, generating ultra-thin fibers that act as electrodes with heightened surface area and porosity. This unique structure enhances the efficiency of charge transfer within batteries. Additionally, electrospinning stands out as a physical technique without a template, employing an intense electric field across the syringe housing the polymer solution and a conducting collector mesh. Sol–gel synthesis represents another widely adopted technique, wherein colloidal suspensions transition into gels, enabling the fabrication of nanomaterial electrodes with regulated porosity and surface properties. Template-assisted synthesis utilizes guides like porous membranes or sacrificial materials to direct nanomaterial growth into specific configurations, allowing for meticulous electrode morphology engineering. Atomic layer deposition provides precise layer-by-layer material deposition, granting fine-tuned control over electrode layer thickness and composition. Hygrothermal synthesis employs high-temperature and high-pressure aqueous solutions to cultivate nanomaterials, offering a straightforward yet powerful means to generate electrodes with desired nanostructures and compositions. These fabrication methods empower researchers to tailor nanomaterial electrodes with optimized attributes for improved battery performance, encompassing enhanced energy density, cycling stability, and rate capability. Table 9.2 provides an overview of various fabrication techniques used for producing nanomaterial electrodes, highlighting their respective advantages, limitations, and examples of materials commonly produced using each technique. The selection of fabrication techniques relies on aspects like desired nanostructure, properties of the material, scalability, and intended use.

Characterization techniques are essential for understanding the structure and electrochemical properties of nanomaterial electrodes, crucial for improving battery performance. Techniques like Scanning Electron Microscopy provide detailed imaging, revealing surface features such as morphology and particle size. Transmission Electron Microscopy offers insights into nanoscale architecture, including crystalline structure and internal morphology, vital for understanding nanomaterial behavior in batteries. X-ray Diffraction helps identify crystal configurations within electrode materials, linking structural changes to battery performance. Electrochemical methods like Electrochemical Impedance Spectroscopy and Cyclic Voltammetry analyze electron exchange and the speed of chemical reactions at the boundary between the electrolyte and electrode. Energy Dispersive X-ray Spectroscopy and X-ray Photoelectron Spectroscopy examine elemental composition and surface chemistry, uncovering factors influencing battery efficacy. Together, these techniques provide a comprehensive understanding of nanomaterial electrode properties, guiding advancements in battery technology [76–78].

9.5.2 Challenges and Future Perspectives of Using Nanomaterials in Electrodes

The advancement in battery performance owes much to discovery of innovative materials used for electrodes, and storage mechanisms. Notably, deliberate creation

Table 9.2 Fabrication techniques for nanomaterial electrodes: advantages, limitations, and examples

Fabrication technique	Advantages	Limitations	Examples of materials produced	References
Chemical vapor deposition (CVD)	• Precise control over film thickness and composition • Scalable for large-scale production	• Requires high temperatures and controlled environments • Limited to certain substrate materials	Graphene, carbon nanotubes, metal oxide thin films	[67]
Electrospinning	• High surface area and porosity • Compatibility with various materials and substrates	• Limited control over fiber alignment and diameter • Prone to process variability	Nanofibrous mats of polymers, carbon nanotube yarns	[68, 69]
Sol–gel synthesis	• Versatile method for producing diverse compositions • Low processing temperatures	• Limited control over nanostructure morphology Requires careful control of reaction conditions	Metal oxide nanoparticles, composite materials	[70, 71]
Hydrothermal synthesis	• Facile and cost-effective synthesis route • Control over particle size and crystallinity	• Limited to hydrophilic precursors • Long processing times	Metal oxide nanoparticles, carbon nanotubes, quantum dots	[72, 73]
Physical vapor deposition (PVD)	• High purity of deposited films • Control over film thickness and stoichiometry	• Limited to certain materials and substrates • High vacuum requirements	Thin films of metals, metal oxides, nitrides	[74, 75]

of nanomaterials has shown potential as a hopeful strategy to surmount numerous inherent difficulties in traditional battery components. Yongming et al. discussed several critical issues encountered in battery technology are investigated, including fluctuations in electrode volume, the development of solid–electrolyte interfaces, as well as the mobility of electrons, ions, and electrode particles [79]. Still finding challenges associated with integrating nanomaterials into battery systems are scrutinized, such as unwanted interactions with electrolytes, limitations in energy density, both volumetric and areal, and the increased expenses resulting from complex manufacturing processes. One of the researchers stated that a primary obstacle involves the necessity to expand nanomaterial synthesis and manufacturing methods to meet the demands of mass production while ensuring cost-effectiveness [80, 81]. Many current processes for fabricating nanomaterials are limited to laboratory environments, leading to concerns regarding their scalability for large-scale commercial use. Ensuring the long-term stability and resilience of nanomaterial electrodes across numerous charge–discharge cycles remains a critical concern. Nanomaterials are prone to degradation mechanisms such as alterations in structure or the formation of particle aggregates, which can result in diminished battery capacity and lifespan. The intricate nanoscale structure of these electrodes may impose constraints on ion diffusion and transport, affecting the battery's capacity to recharge and discharge efficiently and its overall power density. Addressing these challenges will necessitate innovative approaches in nanomaterial design, manufacturing, and integration into battery systems, as well as collaborative research efforts spanning various interdisciplinary fields.

The potential of incorporating nanomaterials into electrodes for battery use shows immense promise, poised to transform energy storage technology. Nanomaterials present unparalleled benefits such as high surface area-to-volume ratios, improved electrical conductivity, and customizable physicochemical properties, enabling the creation of electrodes with superior performance characteristics. With ongoing advancements in nanotechnology and the development of innovative synthesis methods for nanomaterials, coupled with tailored electrode designs, the possibilities for next-generation batteries are vast. Prospects encompass the creation of nanomaterial-based electrodes using enhanced energy concentration, extended cycle life, and accelerated charge/discharge rates, effectively tackling key hurdles in energy storage. Integrating nanomaterials holds potential for realizing adaptable, lightweight, and scalable battery systems applicable across various sectors, from portable gadgets to electric cars, and extensive energy storage grids. As scientific inquiry continues to push forward in both nanomaterial science and battery technology, the outlook for harnessing nanomaterials in electrodes appears highly promising, marking the onset of a new era in sustainable and efficient energy storage solutions. Table 9.4 provides a concise overview of the challenges currently faced in utilizing nanomaterials in electrodes, along with potential future perspectives for addressing these challenges.

Table 9.4 Challenges and future perspectives of using nanomaterials in electrodes

Challenges	Future perspectives
Limited scalability	Development of scalable synthesis methods and manufacturing processes
Cost-effectiveness	Research into cost-effective synthesis routes and recycling methods
Stability and durability	Design of stable nanostructures and protective coatings
Integration with electrolytes	Optimization of interfaces and electrolyte compatibility
Safety and environmental concerns	Investigation of toxicity, recycling, and sustainable production
Performance degradation	Development of novel nanostructures and materials engineering

9.6 Conclusion

This chapter reviews the significant achievements and future potentials in battery technology by investigating different types of battery electrodes, their design and production, the incorporation of nanomaterials, and the use of thermal barrier coatings. Anodes and cathodes composed of lithium, nickel, cobalt, manganese, and other minerals exhibit a variety of electrochemical characteristics that are critical for maximizing battery performance. Improving battery life, charge/discharge speeds, and energy density all depend on how these electrodes are made. Batteries that are more dependable and efficient are the result of advanced production techniques like 3D printing and laser structuring, which are expanding the possibilities of conventional electrode fabrication. Nanomaterial integration has become a game-changing strategy for improving conductivity, surface area, and mechanical stability in battery electrodes. The development of electrodes with improved electrochemical properties made possible by nanotechnology is essential for the production of next-generation batteries with larger capacities and quicker charging periods. Furthermore, one of the most important issues facing battery technology today is heat management, which is addressed by the application of thermal barrier coatings in electrodes. By reducing the possibility of overheating and thermal runaway, these coatings improve battery longevity and safety. Overall, substantial developments in battery technology are being driven by improvements in the design, construction, and materials of electrodes as well as by creative uses of nanomaterials and heat barrier coatings. Not least in the context of electric vehicles and renewable energy integration, these advancements are essential to fulfilling the increasing need for high-performance, safe, and sustainable energy storage systems.

References

1. Kautish P, Lavuri R, Roubaud D, Grebinevych O (2024) Electric vehicles' choice behaviour: an emerging market scenario. J Environ Manag 354:120250. https://doi.org/10.1016/j.jenvman.2024.120250
2. James AT, Kumar G, Pundhir A, Tiwari S, Sharma R, James J (2022) Identification and evaluation of barriers in implementation of electric mobility in India. Res Transp Bus Manag 43:100757. https://doi.org/10.1016/j.rtbm.2021.100757
3. Chang SY, Huang J, Chaveste MR, Lurmann FW, Eisinger DS, Mukherjee AD, Erdakos GB, Alexander M, Knipping E (2023) Electric vehicle fleet penetration helps address inequalities in air quality and improves environmental justice. Commun Earth Environ 4(1):135. https://doi.org/10.1038/s43247-023-00799-1
4. Tsurumaki A, Brutti S, Greco G, Navarra MA (2024) Closed battery systems. Emerging battery technologies to boost the clean energy transition, p 173. https://doi.org/10.1007/978-3-031-48359-2_10
5. Selvi K, Mangai KA, Lett A, Fatimah I, Sagadevan S (2024) Exploring the electrode materials for high-performance lithium-ion batteries for energy storage application. J Energy Storage 92:112208. https://doi.org/10.1016/j.est.2024.112208
6. Subramanian M, Solomon JM, Raja V, Arputharaj BS, Shaik S, Saleel CA, Alwetaishi M, Cuce E (2024) Experimental studies and comprehensive computational investigations on composites-based phase change material for battery thermal management systems in electric vehicles. J Energy Storage 82:110471. https://doi.org/10.1016/j.est.2024.110471
7. Pelletier-Villeneuve B, Schougaard SB (2024) Recent developments in high-power Li-ion battery electrode architecture and active materials: the fast-charging challenge. Curr Opin Electrochem 45:101521. https://doi.org/10.1016/j.coelec.2024.101521
8. Beyers I, Bensmann A, Hanke-Rauschenbach R (2023) Ragone plots revisited: a review of methodology and application across energy storage technologies. J Energy Storage 73:109097. https://doi.org/10.1016/j.est.2023.109097
9. Verma A, Singh A, Colclasure A (2024) On the impact of mechanics on electrochemistry of lithium-ion battery anodes. JOM 76(3):1171–1179. https://doi.org/10.1007/s11837-023-06274-8
10. Dharani AP, Hariharan G, Raja M, Rajeshwaran P, Kumaran S (2023) Improving the performance of Mn_2O_3 electrodes by co substitution for super capacitor applications. J Mater Sci: Mater Electron 34(28):1959. https://doi.org/10.1007/s10854-023-11368-9
11. Bhaskar K, Solomon JM, Sathyamurthy R, Anaimuthu S, Vinayagam NK (2020) A review on PEM fuel cells used for automotive applications, models and hydrogen storage for hybrid electric fuel cell vehicle. https://doi.org/10.4271/2020-01-5173
12. Khan FM, Rasul MG, Sayem AS, Mandal N (2023) Maximizing energy density of lithium-ion batteries for electric vehicles: a critical review. Energy Rep 9:11–21. https://doi.org/10.1016/j.egyr.2023.08.069
13. Tajik M, Makui A, Tosarkani BM (2023) Sustainable cathode material selection in lithium-ion batteries using a novel hybrid multi-criteria decision-making. J Energy Storage 66:107089. https://doi.org/10.1016/j.est.2023.107089
14. Zhou H, Alujjage AS, Terese M, Fear C, Joshi T, Rikka VR, Jeevarajan JA, Mukherjee PP (2024) Effect of fast charging on degradation and safety characteristics of lithium-ion batteries with $LiNi_xCo_yMn_zAl_{1-x-y-z}O_2$ cathodes. Chem Eng J 492:152181. https://doi.org/10.1016/j.cej.2024.152181
15. Subramanian M, Ağbulut Ü, Pachamuthu S, Sathanandam S, Solomon JM, Arputharaj BS, Raja V, Rajendran P, Asif M (2024) Computational and experimental investigations on liquid-based battery thermal management systems for electric vehicle applications under various discharge rates with different flow speeds. J Energy Storage 91:111757. https://doi.org/10.1016/j.est.2024.111757

16. Huang Y-Y, Lin L-Y, Li X (2019) Efficient battery supercapacitor hybrid devices with quaternary metal oxide electrodes based on nickel and cobalt. J Energy Storage 25:100826. https://doi.org/10.1016/j.est.2019.100826
17. Liu W, Placke T, Chau KT (2022) Overview of batteries and battery management for electric vehicles. Energy Rep 8:4058–4084. https://doi.org/10.1016/j.egyr.2022.03.016
18. Kim JG, Son B, Mukherjee S, Schuppert N, Bates A, Kwon O, Choi MJ, Chung HY, Park S (2015) A review of lithium and non-lithium based solid state batteries. J Power Sources 282:299–322. https://doi.org/10.1016/j.jpowsour.2015.02.054
19. Salgado RM, Danzi F, Oliveira JE, El-Azab A, Camanho PP, Braga MH (2021) The latest trends in electric vehicles batteries. Molecules 26(11):3188. https://doi.org/10.3390/molecules26113188
20. Balasubramaniam S, Mohanty A, Balasingam SK, Kim SJ, Ramadoss A (2020) Comprehensive insight into the mechanism, material selection and performance evaluation of supercapatteries. Nano-Micro Lett 12:85. https://doi.org/10.1007/s40820-020-0413-7
21. Entwistle J, Ge R, Pardikar K, Smith R, Cumming D (2022) Carbon binder domain networks and electrical conductivity in lithium-ion battery electrodes: a critical review. Renew Sustain Energy Rev 166:112624. https://doi.org/10.1016/j.rser.2022.112624
22. Gómez-Cámer JL, Bünzli C, Hantel MM, Poux T, Novák P (2016) On the correlation between electrode expansion and cycling stability of graphite/Si electrodes for Li-ion batteries. Carbon 105:42–51. https://doi.org/10.1016/j.carbon.2016.04.022
23. Liu P, Sherman E, Jacobsen A (2009) Design and fabrication of multifunctional structural batteries. J Power Sources 189:646–650. https://doi.org/10.1016/j.jpowsour.2008.09.082
24. Liu X, Zeng Y, Yuan W, Zhang G, Zheng H, Chen Z (2024) Advances in multi-scale design and fabrication processes for thick electrodes in lithium-ion batteries. Energy Rev 3:100066. https://doi.org/10.1016/j.enrev.2023.100066
25. Kuang Y, Chen C, Kirsch D, Hu L (2019) Thick electrode batteries: principles, opportunities, and challenges. Adv Energy Mater 9:1901457. https://doi.org/10.1002/aenm.201901457
26. Xiao X, Liu P, Verbrugge MW, Haftbaradaran H, Gao H (2011) Improved cycling stability of silicon thin film electrodes through patterning for high energy density lithium batteries. J Power Sources 196:1409–1416. https://doi.org/10.1016/j.jpowsour.2010.08.058
27. Sun J, Wu M, Jiang H, Fan X, Zhao T (2021) Advances in the design and fabrication of high-performance flow battery electrodes for renewable energy storage. Adv Appl Energy 2:100016. https://doi.org/10.1016/j.adapen.2021.100016
28. Zhang C, Xu J, Cao L, Wu Z, Santhanagopalan S (2017) Constitutive behavior and progressive mechanical failure of electrodes in lithium-ion batteries. J Power Sources 357:126–137. https://doi.org/10.1016/j.jpowsour.2017.04.103
29. Wu Y, Wang S, Li H, Chen L, Wu F (2021) Progress in thermal stability of all-solid-state-Li-ion-batteries. InfoMat 3:827–853. https://doi.org/10.1002/inf2.12224
30. Zhang K, Yan S, Wu C, Wang L, Ma C, Ye J, Wu Y (2024) Extended battery compatibility consideration from an electrolyte perspective. Small 2401857. https://doi.org/10.1002/smll.202401857
31. Bupesh Raja VK, Raja I, Kavvampally R (2021) Advancements in battery technologies of electric vehicle. J Phys: Conf Ser 2129:012011. https://doi.org/10.1088/1742-6596/2129/1/012011
32. Sankaran G, Venkatesan S (2021) Standardization of electric vehicle battery pack geometry form factors for passenger car segments in India. J Power Sources 502:230008. https://doi.org/10.1016/j.jpowsour.2021.230008
33. Zhang Q, Liang Q, Liao Q, Yi F, Zheng X, Ma M, Gao F, Zhang Y (2017) Service behavior of multifunctional triboelectric nanogenerators. Adv Mater 29:1606703. https://doi.org/10.1002/adma.201606703
34. Şen M, Özcan M, Eker YR (2024) A review on the lithium-ion battery problems used in electric vehicles. Next Sustain 3:100036. ISSN: 2949-8236. https://doi.org/10.1016/j.nxsust.2024.100036

35. Meng Y, Nie C, Guo W, Liu D, Chen Y, Ju Z, Zhuang Q (2022) Inorganic cathode materials for potassium ion batteries. Mater Today Energy 25:100982. https://doi.org/10.1016/j.mtener.2022.100982
36. Jyoti J, Singh BP, Tripathi SK (2021) Recent advancements in development of different cathode materials for rechargeable lithium ion batteries. J Energy Storage 43:103112. ISSN: 2352-152X. https://doi.org/10.1016/j.est.2021.103112
37. Fergus JW (2010) Recent developments in cathode materials for lithium ion batteries. J Power Sources 195(4):939–954. ISSN: 0378-7753. https://doi.org/10.1016/j.jpowsour.2009.08.089
38. Kaur G, Gates BD (2022) Review—Surface coatings for cathodes in lithium ion batteries: from crystal structures to electrochemical performance. J Electrochem Soc 169:043504. https://doi.org/10.1149/1945-7111/ac60f3
39. Wen L, Wang X, Liu GQ, Luo HZ, Liang J, Dou SX (2018) Novel surface coating strategies for better battery materials. Surf Innov 6:1–2, 13–18. https://doi.org/10.1680/jsuin.17.00056
40. Nisar U, Muralidharan N, Essehli R, Amin R, Belharouak I (2021) Valuation of surface coatings in high-energy density lithium-ion battery cathode materials. Energy Storage Mater 38:309–328. ISSN: 2405-8297. https://doi.org/10.1016/j.ensm.2021.03.015
41. Chen Z, Zhang Q, Liang Q (2022) Carbon-coatings improve performance of Li-ion battery. Nanomaterials 12:1936. https://doi.org/10.3390/nano12111936
42. Guan P, Zhou L, Yu Z, Sun Y, Liu Y, Wu F, Jiang Y, Chu D (2020) Recent progress of surface coating on cathode materials for high-performance lithium-ion batteries. J Energy Chem 43:220–235. ISSN: 2095-4956. https://doi.org/10.1016/j.jechem.2019.08.022
43. Rao GVS, Tsang JC (1974) Electrolysis method of intercalation of layered transition metal dichalcogenides. Mater Res Bull 9(7):921–926
44. Lai X et al (2022) Critical review of life cycle assessment of lithium-ion batteries for electric vehicles: a lifespan perspective. eTransportation 12:100169
45. Singh S, Hasan MR, Sharma P, Narang J (2022) Graphene nanomaterials: the wondering material from synthesis to applications. Sens Int 3:100190
46. Liao C, Li Y, Tjong SC (2018) Graphene nanomaterials: synthesis, biocompatibility, and cytotoxicity. Int J Mol Sci 19(11):3564
47. Anzar N, Hasan R, Tyagi M, Yadav N, Narang J (2020) Carbon nanotube—a review on synthesis, properties and plethora of applications in the field of biomedical science. Sens Int 1:100003
48. Eatemadi A (2014) Carbon nanotubes: properties, synthesis, purification, and medical applications. Nanoscale Res Lett 9(1)
49. Mohamed MM, Othman I, Mohamed RM (2007) Synthesis and characterization of MnO_x/TiO_2 nanoparticles for photocatalytic oxidation of indigo carmine dye. J Photochem Photobiol A Chem 191(2–3):153–161
50. Tian W, Fan X, Yang H, Zhang X (2010) Preparation of MnO_x/TiO_2 composites and their properties for catalytic oxidation of chlorobenzene. J Hazard Mater 177(1–3):887–891
51. Suresha B, Divya GS, Hemanth G, Somashekar HM (2021) Physico-mechanical properties of nano silica-filled epoxy-based mono and hybrid composites for structural applications. SILICON 13(7):2319–2335
52. Islam T, Hasan MM, Awal A, Nurunnabi M, Ahammad AJS (2020) Metal nanoparticles for electrochemical sensing: progress and challenges in the clinical transition of point-of-care testing. Molecules 25(24):5787
53. Köhler JM, Kluitmann JJ, Günther PM (2021) Metal nanoparticles as free-floating electrodes. Encyclopedia (Basel) 1(3): 551–565
54. Divya GS, Keshavamurthy R, Siddaraju C, Murthy KVS (2023) Investigation on sliding wear properties of nano metallic particle reinforced hybrid composites through design of experiments and ANN. J Inst Eng (India) Ser D
55. Huang Y et al (2016) Nanostructured polypyrrole as a flexible electrode material of supercapacitor. Nano Energy 22:422–438
56. Nguyen DN, Yoon H (2016) Recent advances in nanostructured conducting polymers: from synthesis to practical applications. Polymers (Basel) 8(4):118

57. Kurc B, Pigłowska M, Rymaniak Ł, Fuć P (2021) Modern nanocomposites and hybrids as electrode materials used in energy carriers. Nanomaterials (Basel) 11(2):538
58. Godja N-C, Munteanu F-D (2024) Hybrid nanomaterials: a brief overview of versatile solutions for sensor technology in healthcare and environmental applications. Biosensors (Basel) 14(2)
59. Choi D et al (2010) $LiMnPO_4$ nanoplate grown via solid-state reaction in molten hydrocarbon for Li-ion battery cathode. Nano Lett 10(8):2799–2805
60. Poizot P, Laruelle S, Grugeon S, Dupont L, Tarascon J-M (2000) Nano-sized transition-metal oxides as negative-electrode materials for lithium-ion batteries. Nature 407(6803):496–499
61. Jamnik J, Maier J (2003) Nanocrystallinity effects in lithium battery materials. Phys Chem Chem Phys 5(23):5215
62. Kim C, Noh M, Choi M, Cho J, Park B (2005) Critical size of a nano SnO_2 electrode for Li-secondary battery. Chem Mater 17(12):3297–3301
63. Yazvinskaya NN, Lipkin MS, Galushkin NE, Galushkin DN (2022) Research of nanomaterials as electrodes for electrochemical energy storage. Molecules 27(3):837
64. Wang J, Lu J, Hocevar SB, Farias PA, Ogorevc B (2000) Bismuth-coated carbon electrodes for anodic stripping voltammetry. Anal Chem 72(14):3218–3222
65. Jiang L et al (2018) A non-enzymatic nanoceria electrode for non-invasive glucose monitoring. Anal Methods 10(18):2151–2159
66. Munonde TS, Raphulu MC (2024) Review on titanium dioxide nanostructured electrode materials for high-performance lithium batteries. J Energy Storage 78(110064):110064
67. Kumar S, Bhushan P, Bhattacharya S (2018) Fabrication of nanostructures with bottom-up approach and their utility in diagnostics, therapeutics, and others. In: Energy, environment, and sustainability. Springer, Singapore, pp 167–198
68. Li D, Xia Y (2004) Electrospinning of nanofibers: reinventing the wheel? Adv Mater 16(14):1151–1170
69. Beckwith KS, Cooil SP, Wells JW, Sikorski P (2015) Tunable high aspect ratio polymer nanostructures for cell interfaces. Nanoscale 7(18):8438–8450
70. Boumya W et al (2022) Modification strategies of sol–gel carbon ceramic electrodes and their electrochemical applications. Results Chem 4(100623):100623
71. Shaheen I, Ahmad KS, Zequine C, Gupta RK, Thomas AG, Malik MA (2021) Modified sol-gel synthesis of Co_3O_4 nanoparticles using organic template for electrochemical energy storage. Energy (Oxf.) 218(119502):119502
72. Chellappa V, Meenakshisundaram N, Annaraj J, Sagadevan S (2024) Hydrothermal synthesis of MnO_2 nanorods for efficient electrochemical detection of environmental anthropogenic pollutants and nitrobenzene. Inorg Chem Commun 160(112015):112015
73. Tajik S, Beitollahi H (2022) Hydrothermal synthesis of $CuFe_2O_4$ nanoparticles for highly sensitive electrochemical detection of sunset yellow. Food Chem Toxicol 165(113048):113048
74. Sedira S, Mendaci B (2020) Hydrothermal synthesis of spherical carbon nanoparticles (CNPs) for supercapacitor electrodes uses. Mater Renew Sustain Energy 9(1)
75. El-Said WA, Alsulmi A, Alshitari W (2022) Hydrothermal synthesis of Mn_3O_4 nanorods modified indium tin oxide electrode as an efficient nanocatalyst towards direct urea electrooxidation. PLoS ONE 17(8):e0272586
76. Saleh HM, Hassan AI (2022) Withdrawn: a review on synthesis and characterization of nanomaterials for application in cost-effective electrochemical devices. Curr Chin Sci 02
77. Manea F (2014) Electrochemical techniques for characterization and detection application of nanostructured carbon composite. In: Modern electrochemical methods in nano, surface and corrosion science. InTech
78. Rudresh M, Surendra BS, Gurushantha K, Dinamani M, Basavaraju N, Deepa H A, Mehrotra R, Golchha T, Mahesh A, Ruhal A (2023) A critical exploratory investigation on nano-inspired ($La_2Zr_2O_7$ NPs) surface coatings for aircraft icing mitigation and electrochemical studies. Australian J Mech Eng 1–12. https://doi.org/10.1080/14484846.2023.2295109
79. Sun Y, Liu N, Cui Y (2016) Promises and challenges of nanomaterials for lithium-based rechargeable batteries. Nat Energy 1(7):16071

80. Charitidis CA, Georgiou P, Koklioti MA, Trompeta AF, Markakis V (2014) Manufacturing nanomaterials: from research to industry. Manuf Rev (Les Ulis) 1:11
81. Surendra BS, Gurushantha K, Anantharaju KS, Rudresh M, Basavaraju N, Raghavendra N, Jahagirdar AA, Somashekar HM, Murthy HCA (2023) Effective paracetamol sensor activity, thermal barrier coating (TBC), and UV-light-driven photocatalytic studies of Zr_xO_2:$Mg^{2+}{}_{(1-x)}$ nanoparticles. New J Chem 47:3978–3992. https://doi.org/10.1039/D2NJ06049C

Chapter 10
Challenges and Opportunities in Nanomaterials for Energy Storage

D. M. Tejashwini, H. M. Deepa, V. V. Deshmukh, and Yashwanth Venkatraman Naik

Abstract This chapter explores the transformative potential and significant challenges associated with the application of nanotechnology in energy storage systems. It highlights the critical role that nanomaterials play in enhancing the performance, efficiency, and sustainability of various energy storage devices, including batteries, supercapacitors, and fuel cells. The chapter delves into the scalability challenges faced in nanomaterial production, emphasizing the need for automated and scalable synthesis processes to facilitate commercial adoption. Additionally, it addresses the integration challenges of nanomaterials into existing systems, focusing on compatibility and handling protocols within traditional manufacturing processes. Despite these challenges, the chapter underscores the vast opportunities presented by recent advancements in nanomaterials. Innovations such as nanostructured silicon anodes for lithium-ion batteries and carbon nanotubes for supercapacitors have demonstrated significant improvements in energy and power densities, as well as cycle life. The chapter also explores future trends in nanomaterial research, particularly the growing emphasis on sustainability and environmentally friendly production methods. It highlights the importance of incorporating green chemistry principles and life cycle assessments in the design of nanomaterials to minimize their environmental impact. In conclusion, this chapter provides a comprehensive overview of the current state and future potential of nanomaterials in energy storage, emphasizing the balance between overcoming existing challenges and seizing emerging opportunities to drive technological advancements in this field.

Keywords Nanomaterials · Energy storage

D. M. Tejashwini (✉) · H. M. Deepa
Department of Studies in Chemistry, Davangere University, Shivagangothri, Davangere 577007, India
e-mail: tejadm1519@gmail.com

V. V. Deshmukh
Department of Physics, Shri Shivaji Science College, Amravati, Maharashtra 444602, India

Y. V. Naik
Department of Physics, RV Institute of Technology and Management, Bangalore 560076, India

R. Naik et al. (eds.), *Advanced Nanomaterials for Energy Storage Devices*, Nanostructure Science and Technology, https://doi.org/10.1007/978-3-031-74730-4_10

10.1 Introduction

Energy storage technologies are essential to modern energy systems as they improve grid stability and allow for the effective use of energy produced from intermittent renewable sources. These technologies may be roughly divided into four categories: chemical, mechanical, thermal, and electrochemical storage systems. Batteries, supercapacitors and fuel cells are instances of electrochemical energy storage devices. High energy density and rechargeability make lithium-ion (Li-ion) and solid-state batteries very desirable. Li-ion batteries tend to be in electric cars and portable gadgets. Li-ion technological advancements, such as enhanced energy density and cycle life by the creation of novel electrode materials and electrolytes that offer greater ionic conductivity and thermal stability, ultimately contributing to more efficient, durable, and safer battery systems [1]. Compared to conventional Li-ion batteries, solid-state batteries provide increased safety and better energy densities since they use solid electrolytes, which reduce the risk of leakage and dendrite formation [2]. Supercapacitors are a form of electrochemical storage technology recognized for their quick charge–discharge speed and high-power density as shown in Fig. 10.1. Quick energy burst applications, like regenerative brakes in cars, are a perfect fit for devices. The use of nanomaterials, with an emphasis on developments like graphene and carbon nanotubes that boost surface area and electrical conductivity, is to improve the performance of supercapacitors [3]. Importance of nanomaterials in advancing energy storage solutions.

Nanomaterials are revolutionizing energy storage systems by drastically increasing performance parameters including energy density, power density, and

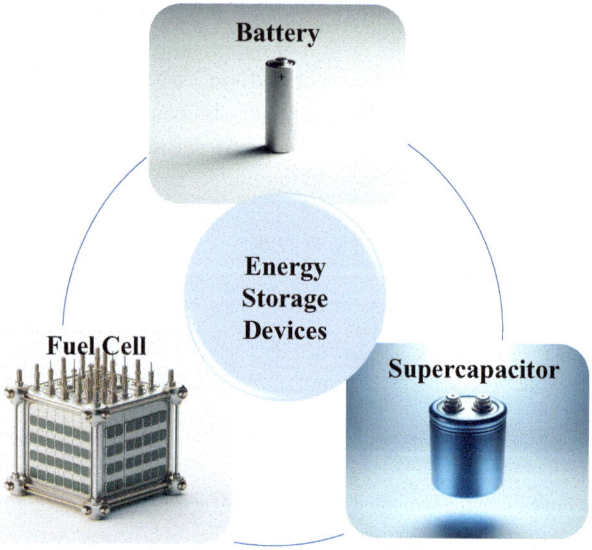

Fig. 10.1 Different types of energy storage device

cycle life. Their special qualities such as their large surface area, improved electrical conductivity, and size-dependent phenomena are essential for maximizing the performance of energy storage systems. Nanomaterials are crucial for increasing the energy and power densities of lithium-ion batteries. The advancement of nanostructured silicon anodes compared to traditional graphite anodes is that silicon has a theoretical capacity that is far larger. However, during charge cycles, silicon experiences volume expansion, which can cause deterioration. Higher energy densities and better cycle stability are provided by nanostructured silicon anodes, highlighting the important role that nanomaterials play in battery performance [4].

Nanomaterials are also very beneficial for supercapacitors. High surface area and great electrical conductivity are provided by graphene-based materials and carbon nanotubes which have revolutionized supercapacitor technology. These nanomaterials improve supercapacitor's energy and power densities, which qualifies them for uses that call for quick cycles of charging and discharging. Nanomaterials facilitate the creation of supercapacitors with markedly enhanced performance metrics such as increased energy storage and quicker charge–discharge rates. Nanomaterials have a significant role in solving ionic conductivity and safety-related problems in the field of solid-state batteries. The significance of nanostructured solid electrolytes in improving the ion conductivity of solid-state batteries in comparison to conventional liquid electrolyte batteries, these nanomaterials enable improved ion transport and enhance the mechanical characteristics of the electrolyte, resulting in safer and more effective solid-state batteries with greater energy densities [5].

The primary objective of this chapter is to provide a comprehensive understanding of how advancements in nanomaterials are driving innovation in energy storage technologies. The chapter aims to illustrate how nanomaterials enhance the performance of energy storage systems. Nanomaterials such as nanostructured silicon anodes in lithium-ion batteries contribute to higher energy densities and improved cycle life. Understanding these impacts is crucial for appreciating the role of nanomaterials in modern energy storage solutions. Another objective is to identify the challenges associated with integrating nanomaterials into existing energy storage systems and explore the opportunities they present. While nanomaterials can significantly enhance the performance of supercapacitors, issues such as scalability and cost-effectiveness need to be addressed. This chapter aims to provide insights into these challenges and discuss potential solutions and advancements. The potential of nanostructured solid electrolytes in solid-state batteries highlights the need for continued innovation in materials science to improve energy storage technologies [6].

- Properties Relevant to Energy Storage

Remarkable qualities of nanomaterials, which dramatically improve performance parameters like energy density, power density, and cycle life, are primarily responsible for their efficacy in energy storage systems. Mechanical strength, electrical conductivity, and surface area are important characteristics related to energy storage. The material's ability to store and release energy is directly impacted by its surface area, making it an essential property. The high surface area-to-volume ratio of nanomaterials improves their capacity to hold more ions or charge carriers. Large surface areas are provided by materials like graphene and nanostructured carbon, which greatly boost the power density and charge storage capacity of supercapacitors [7].

Another essential feature is electrical conductivity. Fast and efficient energy storage requires efficient charge and discharge operations, which are made possible by high electrical conductivity. As compared to bulk silicon, nanostructured silicon anodes in lithium-ion batteries have higher electrical conductivity, which enhances performance and extends battery life [8]. Improved conductivity facilitates quicker energy transfer and more effective functioning by lowering internal resistance. Mechanical strength of nanoparticles is also significant, particularly for preserving structural integrity during charge–discharge cycles. The remarkable mechanical strength of nanomaterials, such as carbon nanotubes and nanowires, contributes to the reduction of degradation and extension of the lifespan of energy storage devices. Compared to traditional electrolytes, nanostructured solid electrolytes in solid-state batteries offer better mechanical stability and durability [9].

10.2 Energy Storage Technologies

Energy storage technologies play a crucial role in controlling the supply and demand of energy for a range of uses, including large-scale grid storage and portable devices. Batteries, supercapacitors, and novel approaches that present particular advantages and difficulties are the main technologies. The most common energy storage technology is a battery, with lithium-ion batteries dominating the market. Lithium-ion batteries, well-known for their high energy density, extended cycle life, and comparatively low self-discharge rates, are widely utilized in electric cars, portable gadgets, and renewable energy storage systems. Even though they are older, lead-acid batteries are still in use because of their affordability and dependability, especially in backup power and automotive applications. Alternative technologies for batteries, such as nickel-metal hydride and sodium-sulphur batteries, also play crucial roles in specific applications [10].

Supercapacitors store energy differently than batteries since they separate charges electrostatically, enabling quick cycles of charging and discharging. Although they lack a comparable energy density as batteries, they are nevertheless perfect for applications that need short energy bursts, such as regenerative braking systems, power backup systems, and some industrial applications because of their high-power density and extended cycle life. In addition to their longevity, supercapacitors are prized for their temperature tolerance. To overcome the shortcomings of the current storage options, new technologies including solid-state batteries, flow batteries, and hybrid systems as well as conventional batteries and supercapacitors are being developed. Solid-state batteries, on the other hand, replace liquid electrolytes with solid ones, offering better energy densities and enhanced safety [11].

- Role of nanomaterials in enhancing these technologies

Nanomaterials can improve present energy storage systems by boosting performance, efficiency, and lifespan as shown in Fig. 10.2. Nanomaterials greatly increase the

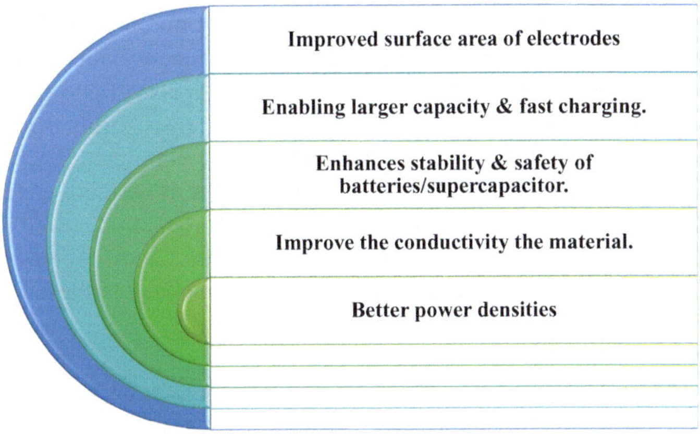

Fig. 10.2 Enhanced properties of energy storage device on implementation of nanomaterials

energy density and charge/discharge rates of batteries. These materials are nanostructured electrodes and nanocomposites. The surface area of electrodes may be increased by nanoparticles of materials like as silicon and titanium dioxide, enabling larger capacity and quicker charging. Although they lessen the chance of deterioration and thermal runaway, nanomaterials also contribute to the stability and safety of batteries. Compared to conventional liquid electrolyte batteries, solid-state batteries, which employ nanomaterials as electrolytes, have higher energy densities and better safety. To improve the electrode material's surface area and conductivity in supercapacitors, nanomaterials like graphene and carbon nanotubes are employed. Faster charge/discharge cycles and noticeably better power densities are the outcomes of this. By enhancing the structural integrity of the electrodes, nanostructured materials also help supercapacitors last longer and have greater mechanical stability. With their ability to improve electrode performance and electrochemical efficiency, nanomaterials are enabling innovative storage technologies like flow batteries [5].

10.3 Challenges in Nanomaterials for Energy Storage

Scalability and repeatability are the primary challenges in the synthesis and manufacturing of nanomaterials for energy storage. One of the biggest challenges is producing on a huge scale while keeping quality constant. The production of nanomaterials, including nanostructured electrodes, frequently results in challenges in managing particle size and homogeneity, which can harm performance [12]. Differences in electrical conductivity and surface area caused by scaling challenges in the synthesis of graphene-based supercapacitor materials affect overall device efficiency [13]. Furthermore, batch-to-batch variability might result from reproducibility issues in nanomaterial manufacturing, such as uneven synthesis techniques, which can impede the creation of dependable and profitably feasible energy storage systems [14]. The research and development of nanomaterials for energy storage is hampered by the cost and complexity of synthesis processes, which also affects the material's scalability and commercial viability. High production costs result from the need for specialized chemicals and sophisticated equipment, as pointed out, which also point out that complex procedures and expensive synthesis methods like chemical vapour deposition and sol–gel processes raise overall production costs [15]. Highlighted that nanomaterial synthesis often requires precise control of environmental conditions, otherwise it further driving up costs and complicating the process [16].

- Performance and Stability

Energy storage systems benefit greatly from the incorporation of nanomaterials; however, maintaining their stability over time is still a major difficulty. Chemical instability is one of the main mechanisms of deterioration, wherein nanomaterials react with other components or electrolytes and eventually lose some of their functionality. Capacity fading and shorter cycle life in lithium-ion batteries can be caused by the chemical interaction between nanomaterial electrodes and electrolyte solutions [17]. Repetitive cycles of charge and discharge weaken the structural integrity of nanomaterials, which leads to mechanical deterioration. This causes problems like electrode expansion/contraction or particle aggregation, which can seriously harm the material's performance. Mechanical stress from cycling can lead to significant structural changes in nanomaterial electrodes, which can affect the electrode's longevity and performance [18]. Surface deterioration caused by air or moisture interaction might have an impact on a nanomaterial's overall efficacy and electrical conductivity. There is a necessity for stabilization techniques and protective coatings by pointing out that surface oxidation and contamination are significant factors in supercapacitors and batteries performance loss [19].

- Issues with Charge/Discharge Cycles

Energy storage devices have been revolutionized by nanomaterials. Yet, there are several issues with their performance throughout cycles of charge and discharge. Capacity fading, or the gradual reduction in a nanomaterial's ability to store and release energy during repeated cycles, is one important problem. The deterioration of electrode materials is a common cause of this issue, which is frequently exacerbated worse by the expansion and contraction of nanomaterials during cycling. Repeated charge–discharge cycles and mechanical stress can result in particle cracking and loss of electrical contact, which lowers capacity and efficiency [20]. Electrolyte deterioration is a further issue that might arise from side reactions between the electrodes made of nanomaterials and the electrolyte. Performance may be further hampered by insulating layers that develop from this deterioration on the electrode surface. Side effects worsen charge/discharge cycle efficiency and accelerate the general deterioration of energy storage devices [21]. Another issue is rate capability as high charge and discharge rates can cause nanomaterial performance to degrade. High power densities can negatively impact energy storage devices' performance and longevity by causing serious problems such as increased internal resistance and thermal instability [22].

- Integration and Practical Application

The incorporation of nanomaterials into current energy storage systems poses several noteworthy obstacles. A significant concern is scalability. Even while nanomaterials frequently perform better in laboratory conditions, increasing the size of these materials for commercial manufacturing might provide challenges. Due to variations in material characteristics and production costs, synthesis and processing methods that are effective in small-scale production may not translate well to large-scale production [23]. Another difficulty is compatibility with current materials. Electrolytes and electrodes two common materials used in energy storage systems must be successfully combined with nanomaterials. Problems may occur with electrical conductivity and contact stability, both of which are essential for effective operation. It can be difficult to ensure that nanoparticles adhere properly and disperse uniformly inside current systems, and this frequently calls for specialized techniques. The issue of economic viability is also addressed. It may be challenging to defend the use of nanomaterials in commercial goods due to their high manufacturing and processing costs. One of the main obstacles to wider use is maximizing the cost-effectiveness of nanomaterial integration without sacrificing performance [24].

- Performance in Real-World Conditions

Evaluating the performance of nanomaterials in energy storage systems under real-world conditions presents several challenges. Laboratory settings often provide ideal conditions that do not accurately reflect the diverse and harsh environments these materials encounter in practical applications. Environmental stability is a major concern, as nanomaterials can degrade or lose efficiency when exposed to factors such as temperature fluctuations, humidity, and atmospheric conditions. Nanomaterials may exhibit reduced stability and performance when subjected to real-world environmental stresses, which are not always replicated in controlled lab environments [25].

Cycling stability is another critical issue. While nanomaterials often show impressive performance during initial cycles, their effectiveness can diminish over time under actual operational conditions. The repeated charge and discharge cycles in practical settings can lead to significant capacity fade and performance degradation due to factors like electrode material wear and electrolyte degradation [26]. Integration with real-world devices also poses challenges. Nanomaterials must be compatible with existing device architectures and manufacturing processes, which may not always be straightforward, highlighting the difficulties in maintaining performance consistency when nanomaterials are integrated into commercial products, where they must function reliably across varying operational conditions and durations [16].

10.4 Opportunities and Advances

The science of nanomaterials is developing quickly, and new structures in energy storage systems are greatly improving functionality and performance. A famous instance of an inventive nanomaterial design is the core–shell structure. These constructions combine the best qualities of two materials by enclosing a core made of one material and a shell made of another. Since the shell protects against deterioration and the core material may function at a high level, core–shell nanoparticles can increase electrochemical stability and capacity. Core–shell nanomaterials greatly improve the energy density and cycle stability of lithium-ion batteries by encasing a high-capacity core within a conductive shell [27]. Another cutting-edge design is represented by hybrid nanomaterials. To make use of complimentary features, these materials combine several nanomaterial types, for illustration, metal oxides, and carbon-based materials. Hybrid structures can offer enhanced conductivity and charge storage capacity.

In comparison to traditional materials, hybrid nanomaterials including carbon nanotube-metal oxide composites perform better in supercapacitors, offering greater densities of energy and power. Furthermore, there has been a lot of promise demonstrated by nanostructured structures like 2D nanosheets and 1D nanowires. High-performance energy storage systems require higher surface area and electrical conductivity, which these designs may provide. 1D nanowires greatly improve charge transfer and battery performance overall, resulting in increased efficiency and quicker charging times [28].

- Advances in Nanomaterial Synthesis Techniques

The performance, scalability, and adaptability of nanomaterials utilized in a variety of applications have been greatly improved by recent developments in nanomaterial fabrication processes. The creation of bottom-up synthesis techniques, which provide exact control over nanomaterial size, shape, and composition, is one noteworthy advancement. CVD, or chemical vapour deposition, is one well-known example. Advanced electronics and energy storage devices require high-quality graphene and other 2D materials with good uniformity and large-scale application, which may be produced through the refinement of CVD [29]. The development of solvothermal and hydrothermal techniques is another noteworthy advancement. These methods include the synthesis of nanomaterials in a solvent under high-temperature and high-pressure conditions. Complex nanostructures, such as hierarchical porous materials, may be created using solvent-free processes, and these materials are advantageous for energy storage applications due to their high surface area and enhanced electrochemical properties [30].

- Enhanced Performance Metrics

The energy storage system's capacity, energy density, and power density have been significantly improved as a result of recent developments in nanomaterials. The quantity of energy that a device can store, termed its ability to store it, has improved significantly as a result of advances in nanomaterial design. As compared to conventional graphite electrodes, the capacity of lithium-ion batteries may be significantly increased by using high-capacity nanostructured electrodes, such as silicon-based anodes, by up to 50% [31]. The ability to store energy per unit of mass or volume, or energy density, has also increased dramatically. More advanced nanomaterials with greater energy storage capacity include nanocomposites and core–shell architectures. By fusing conductive polymers and metal oxides, nanocomposite materials can achieve a tremendous increase in energy density, improving supercapacitors and batteries' performance by over 30% [32]. Power density is a measure of how rapidly energy can be provided or absorbed and has increased as a result of the creation of nanomaterials with higher electrical conductivity and optimized nanostructures. The remarkable electrical conductivity of nanomaterials like graphene and carbon nanotubes enhances supercapacitor's power density and speeds up their charge–discharge cycles. These materials are perfect for high-power applications considering they can attain power densities that are many times greater than those of traditional materials [33].

- Advances in Cycle Life and Efficiency

Modern advancements in nanomaterials have overcome a few of the main shortcomings of traditional technology, greatly extending the cycle life and efficiency of the energy storage system. The number of charge–discharge cycles, a battery, or supercapacitor can sustain before its capacity substantially deteriorates can be improved through innovative nanomaterial designs and cycle life. The addition of nanostructured materials to electrode formulations, such as silicon nanoparticles and carbon nanotubes, has considerably improved cycle stability. Their research indicates that silicon anodes coated with a protective carbon layer might withstand up to 1000 cycles of charge and discharge without experiencing any reduction in capacity. Ordinary graphite anodes, on the other hand, typically show discernible degradation after a reduced number of cycles [34]. Efficiency, which includes both energy and power efficiency, has additionally made substantial progress. The total efficiency of energy conversion and storage has been improved by developments in the synthesis and design of nanomaterials. Supercapacitor's enhanced charge storage and quicker charge–discharge rates are achieved by using nanostructured electrodes with large surface areas and optimized porosity, which boosts the devices' energy efficiency by more than 20% when compared to traditional designs [35].

Enhancement of cycle life and efficiency has been performed possible by stabilization techniques. Protective coatings for nanomaterials and advanced electrolyte additives effectively mitigate issues like electrolyte degradation and electrode material breakdown, further extend cycle life, and maintain high efficiency [36].

- Sustainability and Environmental Impact

Concerning the synthesis and disposal of nanomaterials, in particular, sustainability and environmental effects must be taken into consideration during the development phase. Sustainable synthesis methods are developing to reduce damage to the environment and use of resources. One noteworthy technique is green chemistry, which stays away from hazardous solvents and makes use of renewable, non-toxic ingredients as shown in Fig. 10.3. When compared to conventional methods that use hazardous chemicals, the use of plant-derived precursors and aqueous solutions in the production of metal nanoparticles has greatly reduced the environmental impact [37]. Biological synthesis is becoming more popular due to its advantages for sustainability. This technique offers a more environmentally friendly substitute for traditional chemical methods by using living organisms or extracts to create nanomaterials. When using microorganisms or plant extracts to create nanoparticles, the number of dangerous byproducts is decreased, and the material's performance and characteristics are improved for uses like energy storage and catalysis [38]. Nanomaterial recycling and disposal are essential for reducing their negative effects on the environment. Technological innovations in recycling are being developed to effectively manage waste nanomaterials, innovative methods for recovering and reusing nanoparticles from decommissioned electronics that lessen the need to extract raw materials and help prevent environmental contamination [39].

Reusing and recycling nanomaterials is essential to improving technological sustainability and lessening environmental impact. Recovering valuable materials from items that have reached the end of their useful life and repurposing them for new uses is known as recycling nanomaterials. Chemical recycling, which entails disassembling nanomaterials into their component parts or compounds for further usage, is one efficient strategy. Nanomaterial recycling also involves mechanical separation processes. Using this technique, nanoparticles are physically separated from other waste product constituents. Technological developments effectively separate metallic nanoparticles from used batteries so they may be used again in new energy storage systems. Another interesting approach to recycling is nanomaterial regeneration. Regeneration is the process of giving nanomaterials their original functionality back so they may be utilized again in other applications [40].

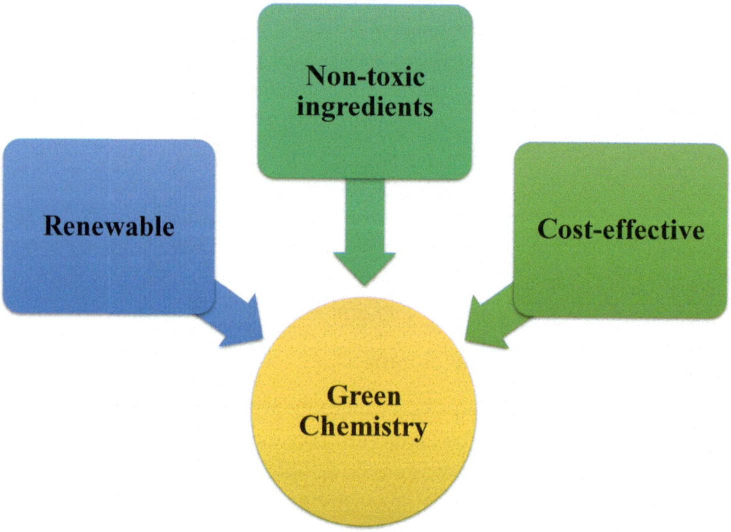

Fig. 10.3 Properties of green chemistry

10.5 Case Studies and Recent Developments

- High-Performance Nanomaterials for Batteries

Significant advancements in lithium-ion and solid-state battery technology have been made recently thanks to developments in nanomaterials, improving key performance indicators including energy density, capacity, and safety. The creation of high-performance nanomaterials has proven essential for lithium-ion batteries. Compared to conventional graphite anodes, silicon-based anodes with nanostructured designs such as silicon nanowires offer much higher capacities and cycle stability. Their silicon nanowires were able to sustain over 1000 cycles with little degradation and reach a capacity of over 2000 mAh/g [41]. Novel nanomaterials have solved issues with interface stability and solid electrolytes in solid-state batteries. A new lithium-ion conducting ceramic nanocomposite electrolyte with nanostructured polymer matrices integrated [42] layered double hydroxides (LDHs) are another instance of an advancement in solid-state battery technology. Cathode materials based on LDH have outstanding rate and capacity capabilities. These materials redefined the performance standard for solid-state batteries with their remarkable energy densities and cycle lives [43].

- Nanomaterials in Supercapacitors

The latest advances in nanomaterials have enhanced supercapacitors' energy density, power density, and cycle stability, resulting in notable gains in their performance. Carbon-based nanomaterials such as carbon nanotubes (CNTs) and graphene have been at the forefront of these discoveries. Graphene aerogel-integrated supercapacitor electrodes have much better specific surface area and conductivity. This innovation has allowed supercapacitors to attain some of the greatest energy and power densities ever reported for these sorts of devices, up to 200 Wh/kg and 20 kW/kg, respectively. Transition metal oxides and conductive polymers have also made substantial contributions. It has been reported that manganese oxide and polyaniline were combined to create a nanostructured composite. This hybrid is made possible by the combined effects of the metal oxide and conductive polymer [44]. The combined actions of the conductive polymer and metal oxide provide this hybrid material with remarkable cycle stability and capacitance. This composite can sustain good performance for 10,000 charge–discharge cycles and reach specific capacitances of above 1000 F/g [45].

- Emerging Technologies

Novel fuel cell technologies and hybrid systems that promise improved performance and efficiency have been developed as a result of recent breakthroughs in nanomaterials. One such instance is fuel cells made of nanomaterials. With nanocatalysts, which increase catalytic activity and lower costs, researchers have made great strides towards improving fuel cell efficiency. Proton exchange membrane fuel cells (PEMFCs) use a platinum-based nanocatalyst supported on graphene sheets that shows remarkable activity for oxygen reduction processes. When compared to traditional platinum catalysts, this invention significantly increased fuel cell performance and durability. Fuel cell hybrid systems, that combine them with other energy storage technologies, are also becoming more and more popular. A hybrid system combines a supercapacitor with a fuel cell based on nanomaterials. Their method increases the efficiency of energy storage and conversion by using a high-surface-area carbon electrode with a nanostructured cobalt oxide catalyst. The shortcomings of conventional energy storage and conversion techniques were addressed by this hybrid system, which showed increased power density and energy efficiency. Furthermore, the performance of solid oxide fuel cells (SOFCs) improved by nanomaterials appears encouraging. The use of nanostructured ceria-based electrolytes in SOFCs was investigated by Kumar et al., and this led to a considerable improvement in ionic conductivity and overall cell performance. Their research demonstrates how these improvements in nanomaterials can result in SOFCs with increased efficiency and lower operating temperatures [46].

10.6 Future Directions and Research Opportunities

Research on nanomaterials is progressing quickly such that novel materials and techniques that have the potential to transform several industries are being highlighted by rising trends. The advancement of two-dimensional (2D) materials beyond graphene is one noteworthy tendency. MoS2 and WS2 are two examples of transition metal dichalcogenides (TMDs) that are gaining popularity due to their remarkable optical, electrical, and catalytic qualities. TMDs are excellent choices for use in flexible electronics, photodetectors, and energy storage devices due to their huge surface areas and strong carrier mobility [47]. The potential for next-generation electrical devices is advanced by their study, which investigates innovative synthesis procedures to create high-quality TMDs with regulated attributes.

The application of biodegradable nanoparticles is another new trend. Research on nanoparticles that can safely decompose after usage is being driven by the demand for ecologically friendly products. Biodegradable nanoparticles are based on silk fibroin for environmental sensing and medication delivery. These materials have the benefit of coming from natural sources, which can reduce their negative effects on the environment while yet offering high functionality and regulated release [48]. According to their research, there are several potential uses for silk fibroin nanoparticles, including environmental monitoring and medicinal implantation. Emerging developments in nanomaterial research have the potential to spur innovation and create new avenues for study in a variety of fields, including environmental science, electronics, and medicine.

- Interdisciplinary Approaches

The assimilation of nanotechnology into many scientific and technical sectors is stimulating revolutionary developments in numerous domains. The convergence of materials science and nanotechnology is well-known by using ideas from both domains; scientists are creating sophisticated nanomaterials with specialized qualities to create nanocomposite materials with remarkable mechanical strength and thermal stability, the fusion of nanotechnology and advanced materials science. These materials provide improved performance for applications in the automotive and aerospace sectors by combining nanoscale fillers with polymer matrices. The development of new catalysts and reaction conditions has been rendered feasible by nanotechnology. The application of nanocatalyst for effective chemical transformations is a noteworthy advancement. Selective hydrogenation processes are using nanocatalyst based on platinum. Due to their enhanced surface area and customized electrical characteristics, these nanocatalysts showed greater activity and selectivity when compared to conventional catalysts. The study demonstrates how catalytic processes and commercial applications may be significantly improved by combining nanotechnology and chemistry [49].

- Commercialization and Industry Impact

Validating the technology, scaling it up, and developing commercialization plans are some of the crucial processes in moving nanotechnology from research to market acceptance. The first and most important stage is technology validation, in which the results of the study are rigorously tested and validated. Comprehensive precommercial trials and pilot programs are the first steps towards successful commercialization. Their study emphasizes how, before increasing production, it is crucial to validate nanotechnology-based solutions in practical settings to demonstrate their effectiveness and dependability. The next big problem is scaling up, which involves converting laboratory-scale technologies into production methods that can be sold commercially. Numerous approaches are adopted to increasing the production of nanomaterials, with a focus on the development of scalable synthesis techniques and manufacturing process optimization [50, 51].

10.7 Conclusion

Nanotechnology has enormous potential to revolutionize many sectors, but it also confronts significant obstacles and possibilities. The scalability of nanomaterial manufacturing is one of the main obstacles. To convert discoveries made in the lab into products that can be sold commercially, scaling up synthesis techniques must overcome major obstacles. One major problem is ensuring cost-effectiveness and consistent quality on a broad scale. Their research highlights the need for automation and scalable synthesis process developments in order to solve these problems and promote commercial use. The incorporation of nanomaterials into current systems is a considerable additional hurdle. It might be difficult to integrate nanomaterials into traditional manufacturing and processing processes because of handling protocols and compatibility problems. Nanotechnology offers a lot of potential in spite of these difficulties. One of the most interesting opportunities is innovation in material properties. Recent developments in catalysis and energy storage with nanomaterials have demonstrated notable performance gains. Technological developments in nanomaterials have the potential to greatly improve energy storage systems by increasing their sustainability, effectiveness, and performance. Energy density, power density, cycle life, and other critical areas are among the many areas where nanomaterials provide advances. The creation of battery electrodes using nanostructures is one significant breakthrough. Lithium-ion batteries have higher energy density thanks to nanostructured silicon anodes. During charge–discharge cycles, silicon experiences expansion and contraction due to its large theoretical capacity; however, these problems are lessened by nanostructure. Capacities above 3000 mAh/g may be achieved using nanostructured silicon anodes, which is a considerable increase over typical graphite anodes and improves total energy storage. Advances in nanomaterials also aid supercapacitors. Carbon nanotubes and graphene-based nanomaterials significantly increase the power density and energy density of supercapacitors. Due to their large surface area and superior electrical conductivity, these materials may produce supercapacitors with power densities of up to 20 kW/kg and energy densities of up to 200 Wh/kg. Their study highlights the potential of nanoparticles to provide extended cycle life and quick charge–discharge. Significant progress has been made in the field of nanotechnology, especially when it comes to energy storage devices. The future of this profession will be shaped by a number of important trends and factors. Nanomaterials have proven to offer enormous promise for improving the efficiency of a variety of energy storage devices, such as fuel cells, supercapacitors, and batteries. But in order to reach their full potential, they must take advantage of new possibilities as well as solve existing problems. The growing emphasis on sustainability and environmental effects is one such trend. The creation of sustainable manufacturing techniques and environmentally friendly nanomaterials is starting to take precedence. The need of using low-toxicity and biodegradable materials to reduce the environmental impact of energy storage systems. It will be essential going ahead to incorporate lifespan analyses and green chemistry concepts into the design of nanomaterials. Subsequent investigations will probably concentrate on improving the functionality and performance of systems based on nanomaterials. Further advancements in energy density, power density, and cycle life might be possible with the development of sophisticated nanostructures and hybrid materials.

References

1. Nagaura T, Tozawa K (2023) Lithium-ion batteries: recent advances and future directions. J Power Sources 560:232206
2. Zhang J et al (2022) Solid-state batteries: advancements and challenges. Adv Energy Mater 12(18):2200610
3. Liu Y et al (2023) Nanomaterials for supercapacitors: enhancing energy and power density. Nano Energy 99:107938
4. Liu Y et al (2022) Nanostructured silicon anodes for high-energy lithium-ion batteries: performance and challenges. Adv Energy Mater 12(6):2100782
5. Wang S et al (2023) Graphene-based nanomaterials for supercapacitors: enhancing power and energy densities. Nano Energy 101:107740

6. Zhang J et al (2021) Nanostructured solid electrolytes for solid-state batteries: enhancing ionic conductivity and stability. J Power Sources 509:230265
7. Aluminates N, Harini HV, Tejashwini DM, Ishwarya S, Deshmukh VV, Nagaswarupa HP, Naik R (2024) 8 Battery and fuel cell. Multifunctional inorganic nanomaterials for energy applications, p 120
8. Nagaswarupa HP, Sillanpää ME, Murthy HA, Naik R (eds) (2024) Multifunctional inorganic nanomaterials for energy applications. CRC Press
9. Wan J, Xie J, Mackanic DG, Burke W, Bao Z, Cui Y (2018) Status, promises, and challenges of nanocomposite solid-state electrolytes for safe and high performance lithium batteries. Mater Today Nano 4:1–16
10. Zhang et al (2023) Hybrid energy storage systems: combining strengths for enhanced performance
11. Chen et al (2022) Supercapacitors in modern energy storage: capabilities and challenges
12. Raghavendra N, Nagaswarupa HP, Shashi Shekhar TR, Mylarappa M, Basavaraju N, Surendra BS (2024) Electrochemical sensor and photocatalytic studies of ferrite nanoparticle using different fuels via green approach. Sens Technol 2(1):2336937
13. Harini HV, Nagaswarupa HP (2024) Green carbon-based nanomaterials to environmental remediation. In: Nanotechnology-based sensors for detection of environmental pollution. Elsevier, pp 95–104
14. Abbas HA, Nasr RA, Khalaf A, Al Bawab A, Jamil TS (2020) Photocatalytic degradation of methylene blue dye by fluorite type $Fe_2Zr_{2-x}W_xO_7$ system under visible light irradiation. Ecotoxicol Environ Saf 196:110518
15. Patete JM, Peng X, Koenigsmann C, Xu Y, Karn B, Wong SS (2011) Viable methodologies for the synthesis of high-quality nanostructures. Green Chem 13(3):482–519
16. Gilbertson LM, Zimmerman JB, Plata DL, Hutchison JE, Anastas PT (2015) Designing nanomaterials to maximize performance and minimize undesirable implications guided by the Principles of Green Chemistry. Chem Soc Rev 44(16):5758–5777
17. Zhao H, Yuan W, Liu G (2015) Hierarchical electrode design of high-capacity alloy nanomaterials for lithium-ion batteries. Nano Today 10(2):193–212
18. Osiak M, Geaney H, Armstrong E, O'Dwyer C (2014) Structuring materials for lithium-ion batteries: advancements in nanomaterial structure, composition, and defined assembly on cell performance. J Mater Chem A 2(25):9433–9460
19. Dutta A, Mitra S, Basak M, Banerjee T (2023) A comprehensive review on batteries and supercapacitors: development and challenges since their inception. Energy Storage 5(1):e339
20. Zhang S, Zhao K, Zhu T, Li J (2017) Electrochemomechanical degradation of high-capacity battery electrode materials. Prog Mater Sci 89:479–521
21. Vukajlović N, Milićević D, Dumnić B, Popadić B (2020) Comparative analysis of the supercapacitor influence on lithium battery cycle life in electric vehicle energy storage. J Energy Storage 31:101603
22. Nadeem F, Hussain SS, Tiwari PK, Goswami AK, Ustun TS (2018) Comparative review of energy storage systems, their roles, and impacts on future power systems. IEEE Access 7:4555–4585
23. Bramsiepe C, Sievers S, Seifert T, Stefanidis GD, Vlachos DG, Schnitzer H, Muster B, Brunner C, Sanders JP, Bruins ME, Schembecker G (2012) Low-cost small scale processing technologies for production applications in various environments—mass produced factories. Chem Eng Process: Process Intensif 51:32–52
24. Qu X, Brame J, Li Q, Alvarez PJ (2013) Nanotechnology for a safe and sustainable water supply: enabling integrated water treatment and reuse. Acc Chem Res 46(3):834–843
25. Riaz A, Sarker MR, Saad MHM, Mohamed R (2021) Review on comparison of different energy storage technologies used in micro-energy harvesting, WSNs, low-cost microelectronic devices: challenges and recommendations. Sensors 21(15):5041
26. Wong WP, Kagalkar A, Patel R, Patel P, Dharaskar S, Walvekar R, Khalid M, Gedam VV (2023) Nano-enhanced phase change materials for thermal energy storage: a comprehensive review of recent advancements, applications, and future challenges. J Energy Storage 74:109265

27. Feng HP, Tang L, Zeng GM, Zhou Y, Deng YC, Ren X, Song B, Liang C, Wei MY, Yu JF (2019) Core-shell nanomaterials: applications in energy storage and conversion. Adv Colloid Interface Sci 267:26–46
28. Mallakpour S, Khadem E (2016) Carbon nanotube–metal oxide nanocomposites: fabrication, properties and applications. Chem Eng J 302:344–367
29. Briggs N, Subramanian S, Lin Z, Li X, Zhang X, Zhang K, Xiao K, Geohegan D, Wallace R, Chen LQ, Terrones M, Ebrahimi A, Das S, Redwing J, Hinkle C, Momeni K, van Duin A, Crespi V, Kar S, Robinson JA (2019) A roadmap for electronic grade 2D materials. 2D Mater 6(2):022001
30. Zhou G, Xu L, Hu G, Mai L, Cui Y (2019) Nanowires for electrochemical energy storage. Chem Rev 119(20):11042–11109
31. Liang B, Liu Y, Xu Y (2014) Silicon-based materials as high capacity anodes for next generation lithium ion batteries. J Power Sources 267:469–490
32. Ishaq S, Moussa M, Kanwal F, Ehsan M, Saleem M, Van TN, Losic D (2019) Facile synthesis of ternary graphene nanocomposites with doped metal oxide and conductive polymers as electrode materials for high performance supercapacitors. Sci Rep 9(1):5974
33. Ishwarya S, Sunagar V, Harini HV, Nagaswarupa HP, Naik R (2024) Electrochemical studies of hybrid nanovanadates and nanotungstates. In: Multifunctional inorganic nanomaterials for energy applications. CRC Press, pp 63–79
34. Zhao L, Ding B, Qin XY, Wang Z, Lv W, He YB, Yang QH, Kang F (2022) Revisiting the roles of natural graphite in ongoing lithium-ion batteries. Adv Mater 34(18):2106704
35. Chen S, Xing W, Duan J, Hu X, Qiao SZ (2013) Nanostructured morphology control for efficient supercapacitor electrodes. J Mater Chem A 1(9):2941–2954
36. Liu J, Yue M, Wang S, Zhao Y, Zhang J (2022) A review of performance attenuation and mitigation strategies of lithium-ion batteries. Adv Funct Mater 32(8):2107769
37. Sharma VK et al (2022) Green synthesis of nanoparticles: eco-friendly methods and applications. Environ Chem Lett 20(1):111–128
38. Patel S et al (2023) Biological synthesis of nanomaterials: green and sustainable approaches. Nano Today 45:101431
39. Zhang Y et al (2021) Recycling and disposal of nanomaterials: challenges and solutions. J Clean Prod 293:125869
40. Patel S et al (2023) Regeneration and reuse of nanocatalysts: techniques and applications. Nano Today 45:101321
41. Feng K, Li M, Liu W, Kashkooli AG, Xiao X, Cai M, Chen Z (2018) Silicon-based anodes for lithium-ion batteries: from fundamentals to practical applications. Small 14(8):1702737
42. Luo S, Liu X, Gao L, Deng N, Sun X, Li Y, Zeng Q, Wang H, Cheng B, Kang W (2022) A review on modified polymer composite electrolytes for solid-state lithium batteries. Sustain Energy Fuels 6(22):5019–5044
43. Lawson K (2021) Synthesis of layered metal hydroxide materials containing nickel, cobalt and copper. Doctoral dissertation, Loughborough University
44. Holze R (2017) Metal oxide/conducting polymer hybrids for application in supercapacitors. In: Metal oxides in supercapacitors. Elsevier, pp 219–245
45. Mulik SV, Jadhav SA, Patil PS, Delekar SD (2022) Transition metal oxide-conducting polymer nanocomposites and metal-organic framework-based composites for supercapacitor application. In: Advances in metal oxides and their composites for emerging applications. Elsevier, pp 135–185
46. Kumar A et al (2021) Nanostructured ceria-based electrolytes for enhanced solid oxide fuel cell performance. J Power Sources 514:230162
47. Gao L (2017) Flexible device applications of 2D semiconductors. Small 13(35):1603994
48. Liu Y, Ai K, Lu L (2014) Polydopamine and its derivative materials: synthesis and promising applications in energy, environmental, and biomedical fields. Chem Rev 114(9):5057–5115
49. Senthilkumar A, Kumar M, Samuel MS, Ethiraj S, Shkir M, Chang JH (2024) Recent advancements in carbon/metal-based nano-catalysts for the reduction of CO_2 to value-added products. Chemosphere 143017

50. Patel S et al (2021) Commercialization strategies for nanomedicine: collaborations, IP management, and regulatory pathways. Nano Medicine 16(7):101–113
51. Johnson M, Lee A (2023) Scaling up nanomaterial production: strategies and innovations for commercial viability. Adv Manuf 12(2):100–115

Chapter 11
Future Perspectives and Emerging Trends in Nanomaterials for Energy Storage

Yashwanth Venkatraman Naik

Abstract Nanomaterials have revolutionized the field of energy storage by offering significantly improved ionic transport and electronic conductivity compared to traditional battery and supercapacitor materials. Their ability to occupy all intercalation sites within the particle volume results in high specific capacities and rapid ion diffusion. These characteristics enable nanomaterial-based electrodes to handle high currents, presenting a promising solution for both high-energy and high-power energy storage applications. Despite these advantages, the commercial use of nanomaterials in energy storage devices remains limited, with notable exceptions like multiwall carbon nanotube additives and carbon coatings on silicon particles in lithium-ion battery electrodes. Over decades of research and development, a diverse array of nanomaterials has been created, featuring a wide range of chemical compositions and shapes, from oxides, chalcogenides, and carbides to carbon and elements forming alloys with lithium. This extensive library includes various particle morphologies, such as zero-dimensional (0D) nanoparticles and quantum dots; one-dimensional (1D) nanowires, nanotubes, and nanobelts; two-dimensional (2D) nanoflakes and nanosheets; and three-dimensional (3D) porous nanonetworks. These nanoscale building blocks, when combined with lithium and beyond-lithium ions, are paving the way for advanced energy storage solutions, including wearable and structural energy storage technologies that are unattainable with conventional materials.

Keywords Nanomaterials · Energy storage · Nanowires

11.1 Advances in Nanomaterials for Energy Storage

The success of nanomaterials in energy storage applications is multifaceted. Nanostructuring is increasingly critical in controlling electrochemical performance and leveraging various charge storage mechanisms, such as surface-based ion adsorption,

Y. V. Naik (✉)
Department of Physics, RV Institute of Technology and Management, Bangalore 560076, India
e-mail: naikyashwanthv@gmail.com

R. Naik et al. (eds.), *Advanced Nanomaterials for Energy Storage Devices*,
Nanostructure Science and Technology, https://doi.org/10.1007/978-3-031-74730-4_11

pseudocapacitance, and diffusion-limited intercalation processes [1–6]. The advent of high-performance materials like redox-active transition-metal carbides (MXenes), which exhibit conductivity far surpassing that of conventional carbon-based electrode materials, has paved the way for the design of current collector-free and high-power next-generation energy storage devices [7]. The integration of nanomaterials into hybrid architectures, such as carbon-silicon and carbon–sulfur combinations, along with the development of versatile nanostructuring techniques, addresses challenges related to the large volume changes typical of alloying and conversion materials. These advancements underscore that nanostructured materials and nanoarchitectured electrodes are essential for designing and realizing energy storage devices that are not only high-energy and high-power but also durable and long-lasting [8].

11.2 Emerging Nanomaterials and Their Potential Impacts on Energy Storage Technologies

Nanotechnology stands as one of the most significant scientific and industrial advancements of the twenty-first century, with applications spanning from electronics to medicine, poised to fundamentally alter lifestyles, work, and global economies [9]. Despite its potential, many of the materials and processes used in nanotechnology depend on non-renewable resources and produce hazardous waste. The future of sustainable technology lies in merging nanotechnological advancements with green chemistry principles [10]. This involves synthesizing nanomaterials using environmentally benign methods and natural ingredients. Although some green nanotechnologies have transitioned from the lab to commercial applications, significant challenges remain. Over the past decade, nanomaterials have shown exceptional promise in diverse fields, including energy storage, pharmaceuticals, and innovative manufacturing techniques. The integration of green chemistry and nanotechnology—referred to as green nanotechnology—focuses on designing processes and products that minimize harmful waste [11]. Recent advancements in this field highlight the use of natural resources, safe solvents, and energy-efficient methods for nanomaterial synthesis. Green nanotechnology is crucial for developing sustainable solutions, particularly in energy and environmental applications, and it represents a pivotal area of scientific and technological progress [12].

11.3 Current Status of Nanomaterials for Energy Storage

Nanomaterials possess unique characteristics that make them integral to the development of high-energy storage devices with significant power density. The conversion of energy in these devices occurs through charge transfer and surface chemical reactions, processes that are optimized at the nanoscale. Nanomaterials provide a large surface-to-volume ratio, increasing the contact surface with electrolytes and thereby offering more reaction sites [13]. This enhanced surface area facilitates electrode reactions that are unattainable with micron-sized particles, crucial for controlling the dynamics of energy storage devices. By reducing diffusion lengths, nanomaterials improve the operational dynamics of these devices, resulting in increased charge and discharge rates [14]. Additionally, nanoscale particles enhance electron transport within the electrode, minimizing stress and lattice damage during repeated charge and discharge cycles. This leads to extended service lifetimes for energy storage equipment. Overall, the incorporation of nanomaterials in energy storage technology offers significant improvements in performance, efficiency, and longevity, making them a key focus for future advancements in this field [1].

11.4 Innovative Synthesis Methods

Advanced Synthesis Techniques for Creating Next-Generation Nanomaterials

The development of next-generation nanomaterials relies heavily on advanced synthesis techniques that enable precise control over size, shape, and composition at the nanoscale [15]. Physical synthesis techniques such as mechanical milling, sputtering, and laser ablation are pivotal in this domain. Mechanical milling involves grinding materials into nanoscale powders using high-energy ball mills, which is essential for producing nanomaterials with unique mechanical and structural properties [16]. The sputtering technique, which involves bombarding a target material with high-energy particles to eject atoms that deposit on a substrate, is widely used in creating thin films with controlled thickness and composition. Laser ablation, another physical method, involves irradiating a target material with a high-intensity laser pulse in a liquid medium, resulting in the formation of nanoparticles with diverse compositions and morphologies. These physical methods are complemented by thermal decomposition, arc discharge, and lithography techniques, each contributing to the creation of nanomaterials with specific characteristics for applications in electronics, catalysis, and energy storage [17].

Chemical synthesis techniques also play a crucial role in the production of advanced nanomaterials. The sol–gel method, for instance, involves the transition of a solution system from a liquid “sol” into a solid “gel” phase, allowing for the fabrication of materials with high purity and uniformity [18]. Coprecipitation, micro-emulsion, and molten salt methods are other prominent chemical techniques used to synthesize nanoparticles with precise control over composition and size. The hydrothermal method, which involves crystallizing substances from high-temperature aqueous solutions at high vapor pressures, is particularly effective for producing nanomaterials with unique crystal structures. Soft and hard templating methods, along with atomic layer deposition (ALD) and chemical vapor deposition (CVD), provide advanced capabilities for designing nanostructured materials with tailored properties [19]. Techniques such as electrospinning and spray pyrolysis are also critical for creating nanofibers and nanopowders with specific functionalities. Additionally, chemical bath synthesis and microwave-assisted methods offer green synthesis alternatives that emphasize sustainability and reduced environmental impact. These advanced chemical techniques collectively drive the innovation and application of next-generation nanomaterials in various technological fields [20].

11.5 The Role of Precision and Efficiency in New Synthesis Methods

The advancement of nanomaterial synthesis is largely driven by the demand for precision and efficiency. Precision in synthesis methods ensures that nanomaterials possess uniform size, shape, and composition, which are crucial for their performance in specific applications. Techniques such as atomic layer deposition (ALD) and chemical vapor deposition (CVD) exemplify this precision, allowing for atomic-scale control over the deposition of thin films and nanostructures [21]. These methods enable the fabrication of materials with highly controlled properties, essential for applications in electronics, catalysis, and energy storage. Additionally, lithography techniques, including electron-beam lithography and nanoimprint lithography, offer unparalleled precision in patterning nanoscale features, which is vital for the development of next-generation electronic and photonic devices [22].

Efficiency in synthesis methods is equally important, as it determines the scalability and economic viability of producing nanomaterials. Techniques such as mechanical milling and spray pyrolysis are valued for their ability to produce large quantities of nanomaterials in a relatively short time. These methods are not only efficient but also adaptable to continuous production processes, making them suitable for industrial applications. Furthermore, the hydrothermal method and microwave-assisted synthesis offer efficient routes to produce nanomaterials with high crystallinity and purity, utilizing energy and resources more effectively compared to traditional methods [23]. The integration of green synthesis techniques, such as chemical bath synthesis and the use of bio-based precursors, also highlights the importance of sustainability in modern synthesis approaches. These methods reduce the environmental impact and enhance the overall efficiency of nanomaterial production, aligning with the global trend toward more sustainable and eco-friendly manufacturing processes [24].

11.6 Breakthroughs in Characterization Techniques

Latest Advancements in the Characterization of Nanomaterials

The characterization of nanomaterials has seen remarkable advancements, enabling researchers to gain deeper insights into their structural, electronic, and chemical properties. One of the most significant breakthroughs is the development of high-resolution transmission electron microscopy (HR-TEM) and scanning transmission electron microscopy (STEM), which allow for atomic-scale imaging and analysis. These techniques, equipped with aberration correctors, can now achieve sub-angstrom resolution, providing unprecedented detail of nanomaterial structures and defects. Additionally, advancements in electron energy loss spectroscopy (EELS) and energy-dispersive X-ray spectroscopy (EDS) integrated with TEM and STEM enable detailed elemental and chemical mapping at the nanoscale, offering insights into the composition and electronic states of materials [25].

Another critical advancement is in the field of in situ and operando characterization techniques, which allow for the observation of nanomaterials under working conditions. In situ transmission electron microscopy (TEM) can monitor dynamic processes such as nanoparticle growth, phase transitions, and electrochemical reactions in real time, providing valuable information on the behavior of nanomaterials in their operational environment. Similarly, in situ X-ray diffraction (XRD) and X-ray photoelectron spectroscopy (XPS) have been developed to study the structural and chemical changes during reactions, giving insights into the mechanisms driving material performance [26]. Scanning probe microscopy (SPM) techniques, such as atomic force microscopy (AFM) and scanning tunneling microscopy (STM), have also seen improvements, offering higher resolution and the ability to manipulate individual atoms and molecules on surfaces. These advancements in characterization techniques are crucial for the design and optimization of nanomaterials for various applications, including energy storage, catalysis, and electronics [27].

How These Techniques Contribute to Understanding and Optimizing Nanomaterials

Advanced characterization techniques are pivotal in comprehending and enhancing nanomaterials by offering profound insights into their structural, electronic, and chemical properties. High-resolution transmission electron microscopy (HR-TEM) and scanning transmission electron microscopy (STEM) allow visualization of nanomaterials at atomic levels, revealing defects, grain boundaries, and interfaces. This detailed structural information helps correlate these features with material properties, enabling the optimization of synthesis methods to minimize imperfections and enhance mechanical and electronic characteristics [28]. In situ and operando techniques, such as in situ TEM, X-ray diffraction (XRD), and X-ray photoelectron spectroscopy (XPS), facilitate real-time observation of nanomaterials under operational conditions. This capability is crucial for understanding dynamic processes like electrochemical reactions and phase transitions, leading to the design of more stable and efficient materials. Scanning probe microscopy (SPM) techniques, including atomic force microscopy (AFM) and scanning tunneling microscopy (STM), offer high-resolution imaging and the ability to manipulate individual atoms and molecules, providing insights into surface properties and interactions. These techniques are essential for optimizing the mechanical and electronic properties of nanomaterials for applications in nanomechanics and flexible electronics. By elucidating the atomic structure, chemical stability, and functional performance, these advanced characterization methods enable the precise tailoring of nanomaterials, fostering the development of next-generation materials with superior performance and reliability [29].

11.7 Computational Advances in Energy Storage

State-of-the-art computational methods are at the forefront of modeling energy storage mechanisms, offering detailed insights that bridge the gap between theoretical understanding and practical applications. Techniques like density functional theory (DFT) and molecular dynamics (MD) simulations provide a deep understanding of the electronic structure and dynamic behavior of materials, crucial for optimizing performance in energy storage devices [30]. DFT, for example, allows researchers to predict essential properties such as voltage profiles, ion diffusion pathways, and material stability, while MD simulations offer a time-resolved view of ion transport and structural changes during charge and discharge cycles.

The integration of these theoretical insights into practical applications is increasingly facilitated by machine learning (ML) algorithms and high-throughput computational screening. ML models can analyze vast datasets generated from computational simulations, identifying trends and predicting material properties with remarkable speed and accuracy. This accelerates the discovery and optimization of new materials for batteries, supercapacitors, and other energy storage devices [31]. Moreover, the development of multiscale modeling techniques, which connect atomic-scale simulations to macroscale phenomena, enables a more comprehensive understanding of how microscopic processes influence the overall performance and lifespan of energy storage systems.

These computational advancements are not just theoretical exercises; they are actively shaping the development of next-generation energy storage technologies. By integrating computational models with experimental data, researchers can refine material properties and design strategies in real time, leading to more efficient, durable, and cost-effective energy storage solutions [32]. This synergy between theory and application is crucial for overcoming current limitations in energy storage, paving the way for innovations that could transform the landscape of renewable energy and sustainable technologies.

11.8 Applications in Cutting-Edge Devices

Use of Emerging Nanomaterials in Novel Energy Storage Devices

Emerging nanomaterials are playing a pivotal role in the development of novel energy storage devices, driving significant advancements in performance, efficiency, and versatility. These materials, with their unique properties such as high surface area, tunable porosity, and enhanced electrical conductivity, enable the design of energy storage systems that are not only more powerful but also more durable and faster charging. For instance, the incorporation of 2D materials like graphene and MXenes into supercapacitors has led to devices with unprecedented energy and power densities, bridging the gap between traditional batteries and capacitors [33]. These materials facilitate rapid ion transport and provide robust frameworks for charge storage, significantly boosting the performance of supercapacitors.

In lithium-ion batteries, emerging nanomaterials such as silicon nanowires and transition-metal oxides have been instrumental in overcoming limitations like capacity fading and slow charging times. Silicon, with its high theoretical capacity, is being explored in the form of nanostructures to mitigate the issue of volumetric expansion during cycling, thus extending battery life and enhancing energy density [34]. Similarly, the use of nanostructured sulfur in lithium-sulfur batteries is showing promise in achieving high specific capacities while addressing challenges related to sulfur's insulating nature and polysulfide dissolution.

Moreover, these nanomaterials are enabling the development of flexible and wearable energy storage devices, which are crucial for the next generation of portable electronics. The ability to create bendable and stretchable electrodes using nanomaterials like carbon nanotubes and conductive polymers is opening up new possibilities for integrating energy storage directly into textiles and other flexible substrates. This innovation not only enhances the functionality of wearable devices but also paves the way for energy storage solutions in emerging fields such as bioelectronics and soft robotics. As research progresses, the continued exploration and integration of emerging nanomaterials will be key to realizing the full potential of novel energy storage technologies [35].

11.9 Examples of Recent Advancements in Device Performance and Efficiency

Recent advancements in energy storage devices have demonstrated significant improvements in both performance and efficiency, driven largely by the integration of innovative nanomaterials and cutting-edge fabrication techniques. For example, lithium-sulfur (Li–S) batteries have seen remarkable enhancements through the incorporation of nanostructured carbon–sulfur composites [36]. These advancements have addressed critical issues such as the shuttling effect of polysulfides, leading to higher cycle stability and significantly improved energy densities. Some of the latest Li–S batteries have achieved energy densities exceeding 500 Wh/kg, which is more than double that of traditional lithium-ion batteries, making them strong candidates for applications requiring lightweight and high-capacity power sources [37].

Supercapacitors have also experienced breakthroughs in performance, particularly through the use of graphene and other 2D materials. By leveraging the exceptional conductivity and surface area of these materials, supercapacitors now boast energy densities that approach those of batteries, while maintaining their inherent advantage of rapid charge and discharge cycles. Recent developments have demonstrated devices with energy densities up to 85 Wh/kg, while still supporting power densities as high as 10,000 W/kg, making them ideal for applications requiring both high power and efficiency, such as in electric vehicles and grid stabilization [38].

In the realm of flexible and wearable electronics, there have been significant strides in integrating nanomaterials into stretchable and bendable energy storage devices. For instance, flexible batteries utilizing nanostructured silicon and carbon nanotubes have been developed, offering not only high-energy densities but also the mechanical flexibility needed for wearable technology. These advancements have led to the creation of batteries that can maintain performance under extreme bending or stretching, ensuring reliable power supply for next-generation wearable devices and flexible electronics [39].

Overall, these examples illustrate how recent advancements in nanomaterials and device engineering are pushing the boundaries of energy storage technology, achieving unprecedented levels of performance and efficiency across a range of applications.

11.10 Challenges in Future Developments

Potential Obstacles in the Development and Implementation of New Nanomaterials

Despite the significant advancements in nanomaterials for energy storage, several challenges remain in their development and implementation. One major obstacle is the scalability of synthesis techniques. While many nanomaterials demonstrate exceptional properties in laboratory settings, producing these materials on an industrial scale without compromising their quality remains difficult. Additionally, the cost of raw materials and the complexity of the synthesis processes can hinder widespread adoption [40]. Another challenge is the stability of nanomaterials under operational conditions. For instance, nanomaterials can undergo structural degradation or unwanted chemical reactions during cycling, leading to a loss in performance over time. There are also environmental and health concerns associated with the production and disposal of nanomaterials, particularly in terms of toxicity and the potential for environmental contamination.

11.11 Strategies for Overcoming These Challenges

To address these challenges, a multifaceted approach is required. For scalability, advancing synthesis techniques to make them more cost-effective and adaptable to large-scale production is crucial [41]. This could involve the development of more efficient chemical processes or the utilization of abundant, low-cost raw materials. Enhancing the stability of nanomaterials can be achieved through surface modifications, the development of protective coatings, or by designing hybrid materials that combine the benefits of multiple nanomaterials while mitigating their individual weaknesses. Addressing environmental and health concerns necessitates the adoption of green chemistry principles in the synthesis of nanomaterials, focusing on reducing hazardous byproducts and developing sustainable methods for the lifecycle management of nanomaterials [42]. Furthermore, fostering collaborations between academia, industry, and regulatory bodies can help establish safety guidelines and best practices for the responsible development and deployment of nanomaterials in energy storage technologies. Through these strategies, the path forward can be paved for the broader and more sustainable implementation of nanomaterials in energy storage solutions.

11.12 Opportunities for Innovation

Areas with Significant Potential for Innovation in Energy Storage

The energy storage sector is ripe with opportunities for innovation, particularly in areas where emerging technologies and materials can offer transformative solutions. One key area is the development of solid-state batteries, which promise higher energy densities, improved safety, and longer lifespans compared to traditional liquid electrolyte-based batteries. Innovations in electrolyte materials, such as solid polymers or ceramic-based electrolytes, can play a critical role in making solid-state batteries viable for commercial use. Additionally, the exploration of alternative battery chemistries, such as sodium-ion, magnesium-ion, or aluminum-ion batteries, offers the potential to create more sustainable and cost-effective energy storage solutions by using more abundant materials [43]. Nanomaterials can also be leveraged to enhance electrode performance, enabling faster charging, greater power output, and longer cycle life. Beyond batteries, there is significant potential for innovation in supercapacitors, particularly through the development of nanostructured materials that can achieve both high-energy density and power density. Moreover, the integration of energy storage with renewable energy systems presents a promising area for innovation, where advanced materials and smart grid technologies can work together to optimize energy use and storage.

11.13 Encouraging Interdisciplinary Collaboration to Drive Advancements

Advancing innovation in energy storage requires a concerted effort across multiple disciplines, including materials science, chemistry, physics, engineering, and computer science. Interdisciplinary collaboration is essential for addressing the complex challenges associated with developing next-generation energy storage technologies. For instance, collaboration between materials scientists and chemists can lead to the discovery of new materials with enhanced electrochemical properties, while engineers can work on scaling up these materials for practical applications. Physicists can contribute by developing theoretical models to better understand the fundamental processes occurring within energy storage systems, which can then be tested and validated by experimentalists. Additionally, integrating computational modeling and data analytics with experimental research can accelerate the design and optimization of new materials and devices [44]. Fostering partnerships between academia, industry, and government agencies is also critical for translating research breakthroughs into commercial products. By encouraging interdisciplinary collaboration, the energy storage field can harness diverse expertise and perspectives, leading to more rapid and impactful advancements.

11.14 Conclusion

This chapter has explored the pivotal role of nanomaterials in revolutionizing energy storage technologies. We began by examining the current state of nanomaterials in energy storage, highlighting their superior properties such as enhanced ionic transport, electronic conductivity, and high specific capacities, which position them as promising candidates for next-generation devices. The chapter also delved into emerging nanomaterials and their potential impacts, emphasizing the synergy between nanotechnology and green chemistry to create sustainable energy solutions. Advanced synthesis techniques were discussed, showcasing how precise control over the nanoscale allows for the development of materials with tailored properties. We also reviewed breakthroughs in characterization techniques, which are crucial for understanding and optimizing nanomaterials, and state-of-the-art computational methods that integrate theoretical insights into practical applications. The application of these nanomaterials in cutting-edge energy storage devices was explored, with examples of recent advancements that have significantly improved device performance and efficiency. Finally, we addressed the challenges and potential obstacles in the development and implementation of new nanomaterials, along with strategies to overcome these barriers, and highlighted areas with significant potential for innovation, emphasizing the importance of interdisciplinary collaboration. As the demand for efficient, high-performance, and sustainable energy storage solutions continues to grow, the role of nanomaterials in this field will only become more critical. The future direction of nanomaterials in energy storage will likely involve the continuous refinement of synthesis techniques to produce materials with even more precise structural and compositional control. Innovations in green nanotechnology will play a vital role in ensuring that these advancements are environmentally sustainable. Furthermore, the integration of advanced computational modeling with experimental research will accelerate the discovery and optimization of novel materials, leading to breakthroughs that were previously unimaginable. Interdisciplinary collaboration will be essential in overcoming the remaining challenges, such as scalability, cost-effectiveness, and long-term stability, to bring these technologies from the laboratory to the marketplace. Ultimately, the successful development and implementation of nanomaterials in energy storage will be a cornerstone in achieving a more sustainable and energy-efficient future, transforming not only how we store energy but also how we think about and utilize it across various sectors.

References

1. Gohar O, Khan MZ, Bibi I, Bashir N, Tariq U, Bakhtiar M, Karim MR, Ali F, Hanif MB, Motola M (2024) Nanomaterials for advanced energy applications: recent advancements and future trends. Mater Des 241:112930. https://doi.org/10.1016/j.matdes.2024.112930
2. Yang C, Chandran KR (2023) A critical review of silicon nanowire electrodes and their energy storage capacities in Li-ion cells. RSC Adv 13(6):3947–3957. https://doi.org/10.1039/D2RA07326A
3. Wang L, Zhang X, Li C, Sun XZ, Wang K, Su FY, Liu FY, Ma YW (2022) Recent advances in transition metal chalcogenides for lithium-ion capacitors. Rare Met 41(9):2971–2984. https://doi.org/10.1007/s12598-022-02028-8
4. Chang F, Davies GL (2024) From 0D to 2D: synthesis and bio-application of anisotropic magnetic iron oxide nanomaterials. Prog Mater Sci 101267. https://doi.org/10.1016/j.pmatsci.2024.101267
5. Vijayakumar V, Anothumakkool B, Kurungot S, Winter M, Nair JR (2021) In situ polymerization process: an essential design tool for lithium polymer batteries. Energy Environ Sci 14(5):2708–2788. https://doi.org/10.1039/D0EE03527K
6. Pholauyphon W, Charoen-amornkitt P, Suzuki T, Tsushima S (2024) Guidelines for supercapacitor electrochemical analysis: a comprehensive review of methodologies for finding charge storage mechanisms. J Energy Storage 98:112833. https://doi.org/10.1016/j.est.2024.112833
7. Liu L, Du Z, Wang J, Li M, Bi J, Du H, Ai W (2023) A review of flexible potassium-ion based energy storage devices. Energy Storage Mater 103022. https://doi.org/10.1016/j.ensm.2023.103022
8. Girirajan M, Bojarajan AK, Pulidindi IN, Hui KN, Sangaraju S (2024) An insight into the nanoarchitecture of electrode materials on the performance of supercapacitors. Coord Chem Rev 518:216080. https://doi.org/10.1016/j.ccr.2024.216080
9. Chowdhury AT, Rafa N, Kabir A, Selvakumar PM (2022) Consumer nanoproducts for environment. In: Handbook of consumer nanoproducts. Springer, Singapore, pp 1169–1200. https://doi.org/10.1007/978-981-16-8698-6_67
10. Dutta D, Das BM (2021) Scope of green nanotechnology towards amalgamation of green chemistry for cleaner environment: a review on synthesis and applications of green nanoparticles. Environ Nanotechnol Monit Manag 15:100418. https://doi.org/10.1016/j.enmm.2020.100418
11. Raja RK, Hazir S, Balasubramani G, Sivaprakash G, Obeth ES, Boobalan T, Pugazhendhi A, Raj RH, Arun A (2022) Green nanotechnology for the environment. In: Handbook of microbial

nanotechnology. Academic Press, pp 461–478. https://doi.org/10.1016/B978-0-12-823426-6.00006-1
12. Thakur M, Sharma A, Chandel M, Pathania D (2022) Modern applications and current status of green nanotechnology in environmental industry. In Green functionalized nanomaterials for environmental applications. Elsevier, pp 259–281. https://doi.org/10.1016/B978-0-12-823137-1.00010-5
13. Kumar A, Dutta S, Kim S, Kwon T, Patil SS, Kumari N, Jeevanandham S, Lee IS (2022) Solid-state reaction synthesis of nanoscale materials: strategies and applications. Chem Rev 122(15):12748–12863. https://doi.org/10.1021/acs.chemrev.1c00637
14. Yu L, Zhou X, Lu L, Wu X, Wang F (2020) Recent developments of nanomaterials and nanostructures for high-rate lithium ion batteries. Chemsuschem 13(20):5361–5407. https://doi.org/10.1002/cssc.202001562
15. Baig N, Kammakakam I, Falath W (2021) Nanomaterials: a review of synthesis methods, properties, recent progress, and challenges. Mater Adv 2(6):1821–1871. https://doi.org/10.1039/D0MA00807A
16. Bishoyi SS, Behera SK (2022) Synthesis and structural characterization of nanocrystalline silicon by high energy mechanical milling using Al_2O_3 media. Adv Powder Technol 33(7):103639. https://doi.org/10.1016/j.apt.2022.103639
17. Baranov O, Bazaka K, Belmonte T, Riccardi C, Roman HE, Mohandas M, Xu S, Cvelbar U, Levchenko I (2023) Recent innovations in the technology and applications of low-dimensional CuO nanostructures for sensing, energy and catalysis. Nanoscale Horizons 8(5):568–602. https://doi.org/10.1039/D2NH00546H
18. Borlaf M, Moreno R (2021) Colloidal sol-gel: a powerful low-temperature aqueous synthesis route of nanosized powders and suspensions. Open Ceram 8:100200. https://doi.org/10.1016/j.oceram.2021.100200
19. Parsons GN, Clark RD (2020) Area-selective deposition: fundamentals, applications, and future outlook. Chem Mater 32(12):4920–4953. https://doi.org/10.1021/acs.chemmater.0c00722
20. Pokrajac L, Abbas A, Chrzanowski W, Dias GM, Eggleton BJ, Maguire S, Maine E, Malloy T, Nathwani J, Nazar L, Sips A, Sone J, van den Berg A, Weiss PS, Mitra S (2021) Nanotechnology for a sustainable future: addressing global challenges with the international network4sustainable nanotechnology. ACS Nano. https://doi.org/10.1021/acsnano.1c10919
21. Chen Y, Fang F, Zhang N (2024) Advance in additive manufacturing of 2D materials at the atomic and close-to-atomic scale. npj 2D Mater Appl 8(1):17. https://doi.org/10.1038/s41699-024-00456-x
22. Zhang Y, Yu H, Wang L, Wu X, He J, Huang W, Ouyang C, Chen D, Keshta BE (2024) Advanced lithography materials: from fundamentals to applications. Adv Colloid Interface Sci 103197. https://doi.org/10.1016/j.cis.2024.103197
23. Manjunatha C, Ashoka S, Krishna RH (2021) Microwave-assisted green synthesis of inorganic nanomaterials. In: Green sustainable process for chemical and environmental engineering and science. Elsevier, pp 1–39. https://doi.org/10.1016/B978-0-12-821887-7.00011-2
24. Kim S, Oh T, Lee H, Nam JM (2022) Trends and perspectives in bio-and eco-friendly sustainable nanomaterial delivery systems through biological barriers. Mater Chem Front 6(16):2152–2174. https://doi.org/10.1039/D2QM00039C
25. Xu M, Li A, Gao M, Zhou W (2020) Single-atom electron microscopy for energy-related nanomaterials. J Mater Chem A 8(32):16142–16165. https://doi.org/10.1039/D0TA04918B
26. Morán MB, Mesa CA, Collado L, Villar-Garcia IJ, Oropeza F, García-Tecedor M (2024) Latest advances on in situ and operando X-ray-based techniques in photoelectrocatalytic systems. J Mater Chem A. https://doi.org/10.1039/D4TA03068K
27. Mubarak S, Dhamodharan D, Ghoderao PN, Byun HS (2022) A systematic review on recent advances of metal–organic frameworks-based nanomaterials for electrochemical energy storage and conversion. Coord Chem Rev 471:214741. https://doi.org/10.1016/j.ccr.2022.214741
28. Jiao S, Fu X, Zhang L, Zeng YJ, Huang H (2020) Point-defect-optimized electron distribution for enhanced electrocatalysis: towards the perfection of the imperfections. Nano Today 31:100833. https://doi.org/10.1016/j.nantod.2019.100833

29. Shahzad K, Mardare AI, Hassel AW (2024) Accelerating materials discovery: combinatorial synthesis, high-throughput characterization, and computational advances. Sci Technol Adv Mater: Methods 4(1):2292486. https://doi.org/10.1080/27660400.2023.2292486
30. Adekoya D, Qian S, Gu X, Wen W, Li D, Ma J, Zhang S (2021) DFT-guided design and fabrication of carbon-nitride-based materials for energy storage devices: a review. Nano-Micro Lett 13:1–44. https://doi.org/10.1007/s40820-020-00522-1
31. Jha S, Yen M, Salinas YS, Palmer E, Villafuerte J, Liang H (2023) Machine learning-assisted materials development and device management in batteries and supercapacitors: performance comparison and challenges. J Mater Chem A 11(8):3904–3936. https://doi.org/10.1039/D2TA07148G
32. Odoi-Yorke F, Opoku R, Davis F, Obeng GY (2023) Optimisation of thermal energy storage systems incorporated with phase change materials for sustainable energy supply: a systematic review. Energy Rep 10:2496–2512. https://doi.org/10.1016/j.egyr.2023.09.044
33. Pandey M, Deshmukh K, Raman A, Asok A, Appukuttan S, Suman GR (2024) Prospects of MXene and graphene for energy storage and conversion. Renew Sustain Energy Rev 189:114030. https://doi.org/10.1016/j.rser.2023.114030
34. Qi Y, Wang G, Li S, Liu T, Qiu J, Li H (2020) Recent progress of structural designs of silicon for performance-enhanced lithium-ion batteries. Chem Eng J 397:125380. https://doi.org/10.1016/j.cej.2020.125380
35. Azim MM, Arifutzzaman A, Saidur R, Khandaker MU, Bradley DA (2022) Recent progress in emerging hybrid nanomaterials towards the energy storage and heat transfer applications: a review. J Mol Liq 360:119443. https://doi.org/10.1016/j.molliq.2022.119443
36. Chen H, Dong WD, Xia FJ, Zhang YJ, Yan M, Song JP, Zou W, Liu Y, Hu ZY, Liu J, Li Y, Wang HE, Chen LH, Su BL (2020) Hollow nitrogen-doped carbon/sulfur@ MnO2 nanocomposite with structural and chemical dual-encapsulation for lithium-sulfur battery. Chem Eng J 381:122746. https://doi.org/10.1016/j.cej.2019.122746
37. Yu B, Chen D, Wang Z, Qi F, Zhang X, Wang X, Hu Y, Wang B, Zhang W, Chen Y, He J (2020) Mo2C quantum dots@ graphene functionalized separator toward high-current-density lithium metal anodes for ultrastable Li-S batteries. Chem Eng J 399:125837. https://doi.org/10.1016/j.cej.2020.125837
38. Balali Y, Stegen S (2021) Review of energy storage systems for vehicles based on technology, environmental impacts, and costs. Renew Sustain Energy Rev 135:110185. https://doi.org/10.1016/j.rser.2020.110185
39. Rahmanudin A, Khan Z, Tybrandt K, Kim N (2023) Sustainable stretchable batteries for next-generation wearables. J Mater Chem A 11(42):22718–22736. https://doi.org/10.1039/D3TA03482H
40. He Q, Zhan F, Wang H, Xu W, Wang H, Chen L (2022) Recent progress of industrial preparation of metal–organic frameworks: synthesis strategies and outlook. Mater Today Sustain 17:100104. https://doi.org/10.1016/j.mtsust.2021.100104
41. Rafiei M, Ricardez-Sandoval LA (2020) New frontiers, challenges, and opportunities in integration of design and control for enterprise-wide sustainability. Comput Chem Eng 132:106610. https://doi.org/10.1016/j.compchemeng.2019.106610
42. García-Quintero A, Palencia M (2021) A critical analysis of environmental sustainability metrics applied to green synthesis of nanomaterials and the assessment of environmental risks associated with the nanotechnology. Sci Total Environ 793:148524. https://doi.org/10.1016/j.scitotenv.2021.148524
43. Bashir T, Zhou S, Yang S, Ismail SA, Ali T, Wang H, Zhao J, Gao L (2023) Progress in 3D-MXene electrodes for lithium/sodium/potassium/magnesium/zinc/aluminum-ion batteries. Electrochem Energy Rev 6(1):5. https://doi.org/10.1007/s41918-022-00174-2
44. Pollice R, dos Passos GG, Aldeghi M, Hickman RJ, Krenn M, Lavigne C, Lindner-D'Addario M, Nigam A, Ser CT, Yao Z, Aspuru-Guzik A (2021) Data-driven strategies for accelerated materials design. Acc Chem Res 54(4):849–860. https://doi.org/10.1021/acs.accounts.0c00785

Index

A
Ab Initio Methods, 26
Advanced synthesis strategies, 99
Atomic Force Microscopy (AFM), 111, 112, 120
Atomic Layer Deposition (ALD), 77, 78, 87, 88, 94
Atomic scale modeling, 28
Automated Synthesis, 165

B
Batteries, 1, 3–12, 15, 16, 21, 57, 67, 68, 71, 73, 165–170, 172–175, 178
Battery performance optimization, 116, 120
Battery technology, 139, 152, 155–158
Biological synthesis, 77, 78

C
Carbon-based materials, 1, 12, 16–18
Carbon nanotubes, 165–169, 172, 173, 175, 178
Catalysts in energy storage
  cobalt oxide nanoparticles, 134
  efficiency and longevity, 135
Catalytic activity, 1, 18, 20, 21
Charge diffusion studies, 111, 112, 118, 120
Charge/discharge rates, 1, 8, 10, 11, 21
Charge storage mechanisms, 57, 70, 72
Chemical Bath Deposition (CBD), 77, 78, 90, 94
Chemical composition, 57, 62, 63, 65, 66, 73
Chemical properties, 58
Chemical synthesis methods, 93
Chemical Vapor Deposition (CVD), 77, 78, 87, 94
Coarse grained modeling, 25
Computational design, 25
Computational nanoscience, 26–28
Co-precipitation, 77, 78, 83, 99, 107
Covalent Organic Frameworks (COFs), 99, 104, 105, 108
Crystallographic structure, 57, 62
Current-Voltage (I-V) Characteristics, 111, 112, 118, 120
Cycle life, 1, 3, 4, 6–8, 11, 12, 15–17, 21
Cycle stability, 139, 152
Cyclic voltammetry, 57, 72, 73
Cycling performance, 57

D
Darcy-Forchheimer, 44
Density functional theory, 25
Doping, 111, 112, 114
Durability, 57, 65, 66, 127, 129, 132–134, 137

E
Electric Vehicles (EVs), 111, 112, 114, 139–143, 148, 150, 158
Electrochemical activity, 111
Electrochemical characteristics, 139, 158
Electrochemical Impedance Spectroscopy (EIS), 57, 63, 71, 73, 111, 112, 117, 120
Electrochemical properties, 57, 65, 71
Electrode design, 139, 144, 152, 157

R. Naik et al. (eds.), *Advanced Nanomaterials for Energy Storage Devices*, Nanostructure Science and Technology, https://doi.org/10.1007/978-3-031-74730-4

Electrode development, 139, 141, 144, 145, 148, 156–158
Electrode materials, 111–120
Electrode optimization, 139, 158
Electrode stability, 57
Electrospinning, 77, 89, 99, 103
Emerging storage technologies, 57
Energy density, 1, 3–10, 12, 15–17, 21, 166–169, 172, 173, 175, 178
Energy efficiency, 58, 68, 70, 71, 173, 176
Energy storage, 1, 3–13, 15–21, 165–174, 176–178
Energy storage applications, 77, 114, 119
Energy storage materials, 57, 65, 68, 70, 72
Energy storage performance, 99
Energy storage systems, 139, 148, 150, 158
Energy storage technologies, 139, 157
  chemical, 124, 125, 130, 131
  electrical, 123, 124, 128, 129, 136
  mechanical, 124, 126–129, 131, 136
  thermal, 124, 125
Energy sustainability
  Germany's 2050 targets, 124
  transition to renewables, 123, 124
Environmental impact, 1, 165, 174, 178

F
Fabrication techniques, 146, 155, 156
Finite difference method, 41, 48
Fourier-Transform Infrared Spectroscopy (FTIR), 57, 65, 66
Fuel cells, 1, 6, 7, 18, 21, 165, 166, 175, 176, 178

G
Graphene, 99, 102, 103, 108
Graphene and carbon-based materials
  applications in batteries and supercapacitors, 128–130
Graphene composites, 153
Green chemistry, 165, 173, 174, 178
Green synthesis, 77, 92, 94

H
Heat source/sink, 41, 42, 44, 47, 50
Hot injection method, 104
Hybrid Electric Vehicles (HEVs), 111, 112
Hydrogels, 99, 105, 106, 108
Hydrothermal synthesis, 1, 13, 18, 20, 21, 85, 86, 94, 103

I
In-situ Characterization Techniques, 111–114, 119, 120
Ion diffusion rates, 57, 71
Ion storage mechanisms, 111, 112, 120

K
Kinetic Monte-carlo simulation, 25, 33, 34

L
Large-Scale Energy Storage Systems (ESSs), 111, 112
Laser ablation, 77–80, 90, 93
Layered Double Hydroxides (LDHs), 99–101, 108
Life cycle assessment, 165
Lithium-Ion batteries, 165, 167, 168, 170, 172, 175, 178
  performance aspects, 128
  silicon nanowires, 126–128

M
Machine learning, 25
Manufacturing compatibility, 165, 171, 177
Material composition, 139, 146
Material optimization, 28
Material simulation, 25–31, 37
Material stability, 57
Maxwell fluid, 41, 43, 49, 50
Maxwell nanofluid, 43, 45
MBenes, 99, 105, 108
Mechanical energy storage, 99–105
Mechanical milling, 77, 78, 93
Mechanical properties, 30, 33
Metal-Organic Frameworks (MOFs), 99, 104, 108
Metal oxides, 1, 4, 8, 11–13, 17, 21
Micro-emulsion technique, 77, 78, 83, 84, 93
Microwave-assisted techniques, 77, 91
Molecular dynamics simulations, 25
Molten salt synthesis, 84
Monte-carlo simulation, 33, 34
Morphology, 57, 67–69, 71–73
MXene, 99, 105, 108

N
Nanocomposites, 1, 8, 12, 21, 99, 102, 107
Nanomaterial characterization, 58, 61
Nanomaterials, 1–4, 8–10, 12, 13, 15–21, 25, 28, 165–178

Nanomaterials in electrodes, 139, 141, 150, 155, 157, 158
Nanoscale properties, 1, 21
Nanoscale structures, 29
Nanostructured materials, 139, 140, 150
Nanostructured silicon anodes, 165, 167, 168, 178
Nanostructuring, 111, 112, 114
Nanotechnology, 37, 165, 176–178
  impact on energy storage and conversion, 124
Neutron diffraction, 111, 112, 119, 120
Next-generation energy storage, 57
Non-Newtonian liquids, 43
Novel nanomaterials, 77, 78, 80, 93

O
Oscillatory stretchy sheet, 45, 54
Oxidation states, 57

P
Particle size, 57, 67, 68, 70, 73
Performance, 123, 125, 126, 128–137
Performance enhancement, 151
Perovskites, 99, 102, 104, 108
Physical synthesis methods, 77
Polymer-based nanostructures, 99, 103
Portable devices, 111, 112
Power density, 165–170, 172, 173, 175, 176, 178

Q
Quantum mechanics, 26

R
Raman spectroscopy, 1, 15, 21
Real-time analysis, 113
Rechargeable batteries, 111, 112, 117, 118
Renewable energy technologies, 57

S
Scalability challenges, 165
Scalable synthesis processes, 165, 177
Scanning Electron Microscopy (SEM), 1, 14, 21, 57, 61, 67, 68, 73, 111, 112, 120
Self-assembly, 99, 103
Silicon and titanium dioxide
  role in battery technology, 137
Silicon composites, 139
Silicon nanowires, 123, 124, 126–128, 135, 136
Similarity variables, 41, 45, 46, 54
Sodium-ion batteries
  cost and scalability, 124, 136, 137
  titanium dioxide nanotubes, 123, 130, 131
Sol-gel method, 81, 83
Sol-gel techniques, 1
Solid-state synthesis, 78, 93
Solvothermal synthesis, 103
Sonochemical method, 99, 107
Spray pyrolysis, 77, 89, 90, 94
Sputtering, 77–80, 93
Structural analysis, 57
Structural design, 139
Structural evolution, 111–113, 117
Sulfides, 1, 11–13, 21
Supercapacitors, 1, 3, 6, 8–12, 16, 17, 21, 57, 67, 71, 73, 99–101, 103–106, 165–170, 172, 173, 175, 176, 178
  graphene, 123, 125, 128–130, 136
  performance metrics, 126, 130, 132, 134
Surface area, 1–3, 6, 8–13, 15, 16, 18, 21
Surface functional groups, 57, 66
Surface modification, 111, 112, 114
Surface properties, 28, 29
Sustainability, 57, 165, 173, 174, 178
Sustainable energy solutions, 139
Synthesis techniques, 77, 78, 80, 81, 83, 90–93
System integration, 165, 171

T
Template-assisted synthesis, 77
Thermal barrier coating, 139, 141, 158
Thermal energy storage, 99
Thermodynamic properties, 27
Transmission Electron Microscopy (TEM), 1, 14, 21, 57, 62, 68, 69, 73, 111, 112, 117, 118, 120

U
Unsteady, 47

V
Viscoelastic, 49, 50

W
Waste biomass-derived materials, 99, 106

X

X-ray Absorption Spectroscopy (XAS), 111, 112, 115, 116, 120

X-ray Diffraction (XRD), 1, 15, 21, 57, 65, 111–115, 120

X-ray Photoelectron Spectroscopy (XPS), 111–113, 119

Z

Zinc-air batteries
- catalysts, 125, 133–136
- performance, 134
- durability, 134

The manufacturer's authorised representative in the EU is Springer Nature Customer Service Centre GmbH, Europaplatz 3, 69115 Heidelberg, Germany. If you have any concerns regarding our products, please contact ProductSafety@springernature.com

Printed and bound by CPI Group (UK) Ltd, Croydon, CR0 4YY

01/10/2026

02225945-0018